BURLINGTON COUNTY HEALTH DEPARTMENT
RAPHAEL MEADOW HEALTH CENTER
WOODLANE ROAD
MOUNT HOLLY, NEW JERSEY 08060

FUNDAMENTALS OF FOOD
MICROBIOLOGY

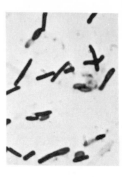

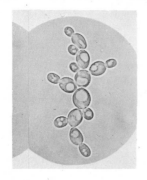

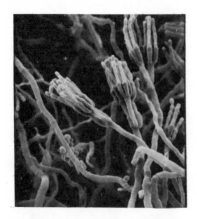

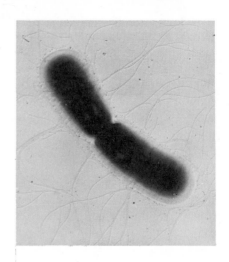

FUNDAMENTALS OF FOOD MICROBIOLOGY

Marion L. Fields

Professor,
University of Missouri,
Columbia, Missouri

AVI PUBLISHING COMPANY, INC.
Westport, Connecticut

Library of Congress Cataloging in Publication Data

Fields, Marion L
    Fundamentals of food microbiology.

    Includes index.
    1. Food—Microbiology.   I.   Title.
QR115.F53   664'.001'576   78-26551
ISBN   0-87055-250-3

Printed in the United States of America

Preface

The contribution of food microbiology to the wholesomeness and pleasure in the food man eats cannot be underrated. And the role of the food microbiologist in the preservation of food and prevention of food spoilage will—without question—remain as dominant in the future as it has in the past.

The author hopes this book will help provide the stimulus and the guidance for microbiological students to enter into the mainstream of very rewarding professional careers. Resolution of the microbiological problems humanity will face in the future will be credited to their achievements.

This textbook started as a manual prepared for a course in food microbiology at the University of Missouri-Columbia. In it, I have attempted to give the fundamentals of food microbiology as taught in our Food Science and Nutrition Course No. 372. The basic principles as outlined in the classic paper of Mossel and Ingram were used as the focal point throughout this book. Because of experience gained in teaching over the years, it seemed best to give the material in basic, condensed form and direct students throughout the book to references which would enlarge their scope and understanding of this important science and which would, simultaneously, encourage and stimulate their interest in this discipline. After questioning students on such a format, they enthusiastically endorsed the condensed organization in presentation of text and the review questions which appear at the end of each chapter. In addition, for greater understanding and accuracy in terminology of the science, a glossary of terms relating to each chapter is given at the end of each chapter.

The final chapter is concerned with methods of how to master and to review food microbiology, and gives 604 questions and answers that test the students' comprehension on what they have gained from the course of study. Outlines of each chapter are provided at the end of the final chapter so that the student can easily see the significance of each subject in relation to the whole body of material in the chapter. These features in this book are unique in this discipline.

The author wishes to acknowledge the permission of authors and publishers who allowed their material to be used in this textbook. Specific acknowledgement to the holder of the copyrighted material is made below. In addition, citations are made within the chapter where such text is presented.

Acknowledgment of Use of Copyright Material

Copyright Holder	Location of Material Used
Elsevier/North-Holland Biomedical Press, Amsterdam	Description of genera of yeasts used in this manual (Chap. 3).
Burgess Publishing Co., Minneapolis, Minn.	Description of genera of molds (Chap. 4).
W. B. Saunders Co., Philadelphia, Penn.	Data in Table 4.2 (Chap. 4).
American Can Co., Barrington, Illinois	Data in Table 9.1 (Chap. 9).
American Chemical Society, Washington, D.C.	Data in Table 11.2 (Chap. 11).
Annals of Applied Biology	Data in Tables 11.4 and 11.5 (Chap. 11).
Institute of Food Technologists, Chicago, Illinois	Data in Table 11.6 (Chap. 11).
American Journal of Public Health, Washington, D.C.	Data in Table 11.7 (Chap. 11).
Institute of Food Technologists, Chicago, Illinois	Data in Table 14.1 (Chap. 14).

I am especially indebted to all researchers and authors who have worked in and written about food microbiology. I have restricted detail as much as possible in the presentation of their knowledge forming the basis of this field today.

Acknowledgement of the assistance of the following persons who have provided pictures for illustrations is made: Dr. Ruth E. Gordon, Dr. M. F. Brown, Dr. H. G. Brotzman, and the Food and Drug Administration.

I am indebted to Thomas Nelson of the University of Missouri-Columbia for the drawings of yeasts in Chapter 3. The assistance of Mrs. Becky Baker and that of my wife are also greatly appreciated.

And, finally, the encouragement and assistance of Dr. D. K. Tressler, Dr. Norman W. Desrosier, Ms. Lucy Long, and Ms. Barbara J. Flouton of the AVI Publishing Company are gratefully acknowledged.

MARION L. FIELDS
Columbia, Missouri

October 1978

Contents

1

Food Microbiology—A Many-Faceted Science

Food microbiology is a science having its roots in antiquity. A food microbiologist is a person who works with foods and studies their microflora.

Food occurs in many forms—red meats, cereals, milk, poultry and poultry products, fruits, vegetables, fish and seafoods, sugar and sugar products. Foods are canned, dehydrated, fermented, frozen, preserved with chemicals and irradiated. All these foods and methods of preservation and marketing procedures make food microbiology a many-faceted and challenging science.

Because of the wide variation in foods, food microbiologists must also have a broad training in food science, as well as microbiology, to be successful. They must know how to isolate, identify and control molds, yeasts and bacteria that are used in food production and/or found in food spoilage.

To have a future in food microbiology, what courses does the student need to get an adequate background to perform his duties in the food industry, universities and government? Obviously, the exact type of work has a bearing upon the subject matter taken. If a student is thinking about industry, a B.S. or M.S. degree will suffice. If he wishes to enter research and has the ambition to become a director of research for a corporation, a Ph.D. is usually required. For these degrees, training in chemistry, physics, food chemistry and food processing are necessary. Courses in beginning microbiology, mycology and/or plant pathology are very helpful. A student may take a beginning and an advanced course in food microbiology. At some schools special courses such as thermobacteriology may be offered. The courses that a food microbiologist and food scientist take are similar, but

1

the food microbiologist has a greater concentration of microbiology courses.

Opportunities are open to qualified microbiologists. As the population increases, as it is predicted to do in the future, competition for available jobs will intensify. Therefore, if you, as a student, want to work in the science of food microbiology, or in any science, you must prepare yourself adequately and strive for excellence. You must work to have more to offer your potential employer than your competition!

DEVELOPMENT OF FOOD MICROBIOLOGY AS A SCIENCE

Food microbiology is a rather new science although some methods of food preservation have been known since early civilization. Important food preservation methods of today are given in Table 1.1. Of the four food preservation methods listed, freezing, dehydration and fermentation can occur in nature. Only the method of canning has no counterpart in nature. Canning was developed, however, before the reasons why it worked were known.

Nicolas Appert began his work on the canning process in 1795 (Desrosier 1977), and it was not until the 1860's that Louis Pasteur made important discoveries in microbiology and food microbiology. One can think of Pasteur as being the father of these sciences because of the following contributions to them:

1. He proved that airborne microorganisms were the cause of growth in nutrient media and that life did not come from spontaneous generation.

2. He suggested that wine be heated, using the controlled heat to prevent wine from becoming sour.

3. He suggested that a wooden rack be placed in barrels of wine undergoing acetification to prevent the acetic acid bacteria from settling to the bottom of the barrel when the barrel was moved. When this happened, the fermentation process was interrupted.

TABLE 1.1. METHODS OF PRESERVING FOODS AND THEIR TIME OF EXISTENCE

Method of Preservation	Time When Method Developed or Became Important Commercially
Freezing	1861, patent issued to Enoch Piper in Maine to freeze fish
	1940, method became important competitor of other consumer-type preserved foods
Dehydration	Oldest and most widely used method of food preservation. Existed in antiquity
Canning	1805, Nicolas Appert given award for inventing process. Process does not exist in nature
Fermentation	Some fermented foods have existed in antiquity

4. He showed that sheep could be vaccinated against anthrax.

5. He made changes in the beer and wine industries to improve their products.

6. He devised the method of heating milk at 61.7°C (143°F) for not less than 30 min to destroy pathogens.

7. He established a method to make chickens immune to chicken cholera.

8. He developed a method for treating rabies.

9. He showed that milk could be soured by inoculating microorganisms from buttermilk or beer, but the milk was unchanged if not inoculated.

10. He isolated bacilli that caused disease of silkworms by developing methods of detecting the disease and ways to prevent it.

It was Pasteur, therefore, who laid the groundwork for the science of food microbiology. After Pasteur came several men who dealt with the microbiology of canning. Because canning is a preservation process that has no counterpart in nature and that has been researched over the years, this will be discussed in the development of food microbiology as a science. Bitting (1937) cited the following researchers in the canning industry.

1. H. L. Russell. In 1895, he applied bacteriology to the spoilage of canned peas in Wisconsin. His recommendation of increasing the process from 10 to 12 min at 110°C (230°F) to 15 min at 116.6°C (242°F) solved the spoilage problem which the industry had.

2. S. C. Prescott and Lyman Underwood. They presented a series of papers about bacterial spoilage of canned products to the canning industry. The first paper, presented in 1896, was entitled "Microorganisms and Sterilization Processes in the Canning Industries."

3. E. W. Duckwall. The first commercial laboratory was established by Duckwall in 1902 and he published "Canning and Preserving with Bacteriological Techniques" in 1905.

4. Branson Barlow. He was the first person to study the importance of thermophilic, sporeforming bacteria that causes flat sour spoilage in corn and pumpkin.

5. W. D. Bigelow. In 1921, Bigelow showed that bacterial spores die logarithmically when subjected to lethal heat.

6. C. O. Ball. In 1923 and 1928, Ball developed mathematical methods of determining the process times for canned foods.

During and after the 1940's, more microbiological research was done on frozen foods. During and after World War II, research on the quality of dehydrated foods was carried out. As new methods and/or equipment have been developed, there has been microbiological research involved in their production by industry and evaluation by governmental agencies. As we shall discuss shortly, food microbiologists have been working on food poisoning problems since A. Gärtner first isolated *Salmonella enteriditis*.

Today, we, in the United States, enjoy the safest and most plentiful food supply in the world. By preventing as much spoilage as we can and by developing new methods to process and preserve food, we create an environment that is conducive to better national health for America.

The development of food microbiology as a science continues today. The food microbiologist works closely with the food scientist in this evolutionary process.

INTERRELATIONSHIPS OF FOOD MICROBIOLOGY WITH OTHER SCIENCES

Much of the information developed in the "basic" or "pure" sciences has application in food microbiology. The scientists who are involved in the basic sciences study objects, substances and phenomena without necessarily thinking of practical application of the knowledge to help solve problems that are involved in everyday life.

Food microbiology is an applied science. In this science, basic scientific principles from mathematics, mycology, plant pathology, bacteriology, chemistry and physics are applied in solving problems involving foods and microorganisms. With this being the case, the student can readily understand the importance of such "pure" science courses as mathematics, chemistry and physics in the course of study for his degree. Examples of some science concepts being applied to microbiology of canning are shown in Table 1.2.

TABLE 1.2. EXAMPLES OF OTHER SCIENCES APPLIED TO MICROBIOLOGY OF CANNING

Science	Concept
Bacteriology and chemistry	Foods with a pH of 4.5 and lower may be processed at 100°C (212°F)
Bacteriology and chemistry	Spores of *Clostridium botulinum* do not germinate below pH 4.8
Physics	Heat flows from hotter to colder environment. This is the second law of thermodynamics
Mathematics	Ball's equation for a simple heating curve is: $B_B = fh \, (\log I - \log g)$
Mathematics	When the area under the lethality curve is equal to the unit area, which is equal to an F value of 1 at 121°C (250°F), there will be 100% kill of the bacterial spores (General Method of Process Calculations)

WHERE AND HOW THE FOOD MICROBIOLOGIST IS EMPLOYED

A. Government. Employment in government can occur at the federal, state, county and city level. Governmental agencies will be setting standards regulating the numbers and kinds of bacteria foods can contain. After these criteria are set, more microbiologists will be needed to check on the purity of foods on the markets. An FDA microbiologist is pictured in Fig. 1.1.

B. Universities. Universities hire scientists in food science departments to teach and to do research on problems in microbiology. At present, there are more opportunities in industry and government than in universities. A university microbiologist is preparing to use an electron microscope in Fig. 1.2.

C. Industry. Companies have their own quality control procedures. These may be more strict than the governmental requirements. Of course, industry must see to it that their products are in compliance with all regulations and laws. This means that industry needs many food microbiologists also. See Fig. 1.3.

D. Employment Bureaus. Students just graduating from college may or may not know about specific employment possibilities in food microbiology and/or food science. The Institute of Food Technologists (IFT) assists students as well as employed workers (who want to make a change)

Courtesy of Food and Drug Administration

FIG. 1.1. IN DALLAS DISTRICT LABORATORY, AN FDA BACTERIOLOGIST CHECKS FOR *E. COLI*
Bacterial growth is observed in tubes and petri dishes.

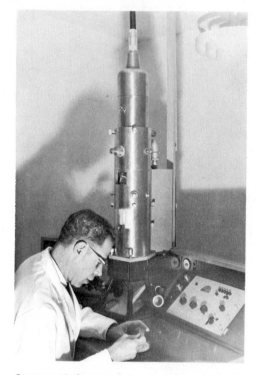

Courtesy of R. Goodman, University of Missouri-Columbia

FIG. 1.2. A UNIVERSITY FOOD MICROBIOLO-
GIST EXAMINES MATERIAL TO BE VIEWED BY
THE ELECTRON MICROSCOPE

to obtain positions. Examples of the type of industrial food manufacturing companies and the types of duties of the food microbiologist are listed in Table 1.3. This table was prepared from material distributed at the IFT Employment Bureau at the 1978 Annual Meeting in Dallas, Texas. One can assume that similar positions will be open in the future and that scientific developments and processes will expand the need for food microbiologists.

The types of duties required in these job descriptions included supervision, quality control, plant sanitation, training programs, dealing with regulatory agencies, trouble-shooting and performing special tests. One company wanted a person to participate in management. These certainly are not all the types of employment that occur in industry but may give you, the student, some idea as to what to expect.

In addition to the IFT Employment Bureau at national meetings, IFT publishes a professional placement service in the back of *Food Technology,*

Courtesy of N. Finley, 7-Up Company, St. Louis

FIG. 1.3.   AN INDUSTRIAL FOOD MICROBIOLO-
GIST EXAMINES FOOD SAMPLES FOR MICRO-
ORGANISMS

a monthly publication. Advertisements can be placed by either the employer or potential employee.

Independent employment agencies are available, also, to help students
obtain employment. Usually the client companies pay all fees and expenses.
Advertisements by these firms are listed in *Food Technology*.

The American Society for Microbiology publishes the *ASM News*
monthly. They also publish positions available and positions wanted in
the employment section of the magazine.

CAREERS OF THE FUTURE

As we will soon discuss, the problems of the future are staggering as the
world population increases. Certainly, food microbiologists, working closely
with food scientists, have much to do. If you, as a student, want to take on
a challenging profession that can help people stay alive and have hope for
the future in the world, then food microbiology is such a profession. Its
importance cannot become less but can only increase with each succeeding
year.

In the career of the future, you probably will be working with single-cell
protein derived from microorganisms. Microorganisms are the greatest and
most efficient producers of protein. Some of these single-cell proteins may
be produced upon conventional substrates where the carbohydrate source
may be starch or a breakdown product of starch.

Cellulose will be the most important carbohydrate source of the future

TABLE 1.3. EXAMPLES OF DUTIES OF FOOD MICROBIOLOGISTS WITH B.S. DEGREES IN SELECTED FOOD INDUSTRIES

Type of Company	Type of Duties of Food Microbiologists[1]
Candy company	Working with vendors to improve sanitation
	Sanitation in processing plant
	Determining quality of incoming supplies
	Training production personnel in personal hygiene
Specialty products	Quality control by bacteriological testing of food products
Chemical and engineering firms	Quality control with good advancement potential
Chocolate manufacturer	Supervision of quality control
	Implementation of new standards
	Plant sanitation
Grocery company	Perform standard bacteriological tests for *Salmonella, Staphylococcus,* etc.
	Test raw materials and finished products
	Duties in heat processing, sterility and shelf-life studies
	Supervisory duties
Soft drink manufacturer	Provide microbiological services in establishing manufacturing procedures
	Quality control programs
	Trouble-shooting and development activities
Meat packer	Development and management of quality control system
	Dealing with regulatory agencies
	Participation in operations management team
Independent testing laboratory	Microbiological analysis of foods

[1] Description of jobs listed at the Institute of Food Technologists Employment Bureau at the 1978 Annual Meeting in Dallas, Texas, on June 4–7.

because it is the most abundant carbon source for the production of single-cell protein. Research using cellulolytic bacteria is being done now and more will be done in the future. There is a possibility of using a combination of microorganisms to produce protein. For example, use of algae to produce the carbohydrate source and use of cellulolytic bacteria and/or molds to produce single-cell protein from the algae are possibilities. Such a system would not use carbohydrates from conventional sources to produce protein.

The limit of our abilities to solve food problems is the limit of our imagination.

THE ROLE OF SCIENTISTS IN FOOD PRODUCTION AND SAFETY

The current world population is 4 billion people. By the end of the next 25 years, the total is projected to reach 7 billion. All of the new individuals will seek the good life. This means more food will be required than ever before in the history of mankind.

Some of the needed increased production can be achieved by increasing the area used for cultivation of crops. The amount of land which can be cultivated is finite. There is a limit, therefore, to this means of increasing food supplies. Another method is to develop new varieties which will yield more per acre. Varieties such as the Green Revolution produced are also linked with fertilizer application. and fertilizers are more expensive than they used to be. Most of the increase in population will be from the poor in developing countries who do not have money for these adjuncts. These problems, unless solved, promise a bleak and gloomy future. It is up to the food microbiologist, the food scientist and all agriculturalists to work together to solve these problems.

Preventing Spoilage

Many foods decay before they are consumed. This occurs in the field (in fruits or vegetables, for example), in marketing channels, or in storage. Methods to reduce spoilage would make more food available for consumption and for export. This implies that a search must be made for ways to preserve chemically fruits and vegetables and that such means should use the least amount of energy. Only disease-resistant varieties should be grown.

Since most humans obtain their protein from cereals, only high-yielding varieties should be used. To save energy in drying grains, safe chemicals to prevent the growth of filamentous fungi should be researched.

All foods should be studied to prevent spoilage. Research which deals with fruits, vegetables and cereals will be of value for more people on a world basis than studies dealing with meat, eggs, seafoods and dairy products since fewer people (on a worldwide basis) have money to purchase these more expensive foods. On the other hand, if more cereal products are available for both man and animals, more food can be produced.

Food Poisoning and Infection

The National Research Council has given its views to the United States Department of Agriculture on the type of research needed to ensure the microbiological safety of foods and made the following recommendations for research in food microbiology:

1. *Salmonella*
   (a) Find faster and simpler methods for detection.
   (b) Develop methods to eliminate *Salmonella* from dried foods.
   (c) Survey foods for the presence of salmonellae.
   (d) Determine how foods become contaminated.
   (e) Determine how salmonellae are pathogenic in animals.
   (f) Determine factors involved in the destruction and prevention of growth of these bacteria.

2. *Staphylococcus*
    (a) Detect new enterotoxin and determine their chemical and im-
         munological properties and the relationship between them.
    (b) Develop better methods of detecting and enumerating enterotoxin
         in man; determine how the toxin is formed and the factors af-
         fecting its formation.
    (c) Determine ways to prevent growth in fermented foods.
    (d) Study the relationship of staphylococcal mastitis to food poi-
         soning.
3. *Clostridium perfringens*
    (a) Research more basic information about the nature and mecha-
         nisms of *C. perfringens* poisoning and the means of controlling
         the bacterium.
    (b) Study ways to detect and quantify the bacterium.
    (c) Identify strains with potential to cause food poisoning.
4. **Mycotoxins**
    (a) Develop methods to detect mycotoxins.
    (b) Study their mode of action in animals.
    (c) Determine their effect on primates.

Of course, these are not all the possible areas for research in food microbiology but this list does give one an idea of what some experts considered to be problems in the United States in 1972. Many times in meeting
a need, technological advances create new ones. One can be sure that there
will be many challenges in the future.

Single-Cell Protein

According to Scrimshaw *et al.* (1975), the role of protein, obtained from
yeast or bacteria, as a supplement to the world supply is now well established. These scientists feel that some form of single-cell protein will ulti-
mately fulfill a major food need for mankind. They state that 10% supplement to the world's food supply could be provided by a fermenter occupying
an area equivalent to about 1.5 square kilometers. An illustration of production on a laboratory scale is found in Fig. 1.4.

Priorities of studies on single-cell protein are based upon the following:

(a) Significance of research in the economics of the overall process.
(b) Importance of the investigation to expansion of direct use in food.
(c) Probability of success of the investigation.

These priorities are based upon industrial production. Significant by its
absence is what could be done on a home basis for developing procedures

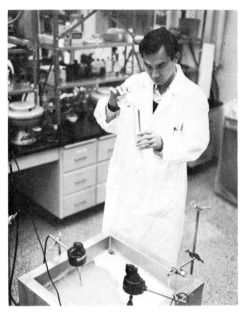

Courtesy of University of Missouri-Columbia

FIG. 1.4. LABORATORY PRODUCTION OF
CANDIDA TROPICALIS FOR STUDIES WITH
SINGLE-CELL PROTEIN

to help poor people use fermentation for enriching their foods. In Fig. 1.5 the research team examines corn muffins made with added protein.

Milner *et al.* (1978) list the following research needs for nonphotosynthetic single-cell protein:

1. Develop fermentation procedures that approach the theoretical yield economically.

2. Devise methods for removal of RNA from the cells.

3. Make more efficient use of proteins within the cells (isolation procedures).

4. Determine interaction of single-cell proteins with other food constituents.

5. Select microorganisms that grow at high temperature to avoid high RNA content and to avoid the need for sterilization and cooling of substrate.

6. Develop mutants which will excrete sulfur-containing amino acids for fortification of both single-cell and plant proteins.

7. Develop more economic methods for harvesting cell mass.

8. Determine better ways to dewater and to dry the cell mass.

Courtesy of University of Missouri-Columbia

FIG. 1.5.   RESEARCH TEAM EXAMINES CORN MUFFINS PAR-
TIALLY MADE FROM SINGLE-CELL PROTEIN

9. Ascertain ways to use cellulose and/or waste materials as substrates.

These are examples of the ever-expanding field of industrial efforts to
increase protein needs for animals and man. It is obvious that more information is needed in food science and food microbiology in such an ex-
panding technology.

THE ROLE OF FOOD SCIENCE PROGRAMS

In the United States

According to Buchanan (1972), the mission of food science problems, as
they involve microbiology, is to give the student a thorough knowledge of
food microbiology so that the students can work on food safety problems.
Of course, the role is to reduce spoilage as much as possible and to develop
techniques for the production of single-cell protein.

In the Developing World

The problems of the developing world are the same as those in the United
States except they are more intensified. Because of the increased population
in these countries, there is a need to produce and preserve as much food
for man as possible.

TRAINING IN FOOD MICROBIOLOGY

Institute of Food Technologists' Recommendations

The Institute of Food Technologists outlines standards for undergraduate education in food science and technology. The minimum requirements for the curriculum and subject matter are given.

The IFT curriculum recommendations dealing with food microbiology contain the following major sections: microorganisms; morphology; culture; physiology and activities in food; inactivation or removal; thermal process evaluation; microorganisms and their properties utilized in processing; microbiological examination; foods and public health standards for foods.

As can be seen by the wide range of the above outline, to become a thoroughly knowledgeable food microbiologist does, indeed, require more than one course of study and more than one year of concerted education in the discipline. This textbook makes its contribution toward that end by covering these broad areas, keeping details to a minimum, and giving relevant references throughout for supplemental study. The individual's interest may thus be extended and concentrated toward chosen objectives. Examples of such objectives include:

To become familiar with the characteristics of important microorganisms of foods.

To learn the techniques of microbiological evaluation of foods.

To be able to discuss intelligently the role of microbes in the production, processing and keeping qualities of foods.

To strive toward the day when the very important subject of foodborne diseases and their control is a familiar subject.

And—perhaps most important of all—to strive for excellence in every facet of the many-faceted science of food microbiology, since in your career as a food microbiologist you will be called upon, numerous times, to make extremely critical decisions that affect the well-being of man and the wholesomeness of his food. Suggested material for a semester course is given in Table 1.4.

FOOD MICROBIOLOGICAL LITERATURE

Textbooks

The first textbook in the United States devoted to the subject of food microbiology (*Microbiology of Foods*) was by Fred W. Tanner. The 2nd edition (and last) was published in 1944 by The Garrard Press, Champaign,

TABLE 1.4.   SUGGESTED MATERIAL FOR A SEMESTER COURSE

Lectures		Laboratories	
Subject Area	Class Periods[1]	Subject Area	Class Periods[2]
Tests for microorganisms in foods	3	Direct counting of bacteria	2
Bacteria important in foods	5	Standard plating techniques	2
Yeasts	3	Study of yeasts	2
Molds	3	Study of molds	2
Factors affecting microbial growth	2	Micrococcaceae	2
Microbiology of fruits and		*Streptococcus* and	
vegetables	2	*Leuconostoc*	2
Microbiology of eggs and poultry	1	*Bacillus, Clostridium*	4
Microbiology of fish and sea foods	1	*Salmonella*	2
Microbiology of red meats	3	Coliforms	2
Microbiology of canned foods	2	*Pseudomonas,*	
Microbiology of milk and dairy		*Flavobacterium,*	
products	4	*Alcaligenes*	2
Microbiology of dehydrated foods	2	Unknowns	17
Chemically preserved foods	2		
Refrigerated foods	2		
Food poisoning	3		
Total	39		39

[1] Based upon three classes per week.
[2] Based upon two laboratory periods per week, each being 2 hours.

Illinois, and became a textbook for courses of food microbiology in the late forties and fifties. Those parts of the book which were based upon principles are still valid today. The book has the following chapter titles: Food Preservation; The Bacteria; Yeasts and Molds; Bacteriology of Water and Sewage; Microbiology of Milk; Bacteriological Milk Analysis; Pasteurization of Milk; Microbiology of Cream and Butter; Microbiology of Cheese; Microbiology of Frozen Desserts, Ice Cream and Similar Products; Microbiology of Concentrated Milks; Fermented Milks; Intestinal Microbiology; Microbiology of Fruit and Fruit Products; Microbiology of Vegetables and Vegetable Products; Microbiology of Tomato Products; Microbiology of Bread; Fermented Foods; Microbiology of Sugar and Sugar Products; Microbiology of Fish and Shellfish; Microbiology of Meat and Meat Products; Microbiology of Egg and Egg Products; Microbiology of Canned Foods; Microbiology of Miscellaneous Food Products; Microbiological Methods of Assaying Foods for Vitamins and Culture Media. As can be seen by the chapter titles, the book was very comprehensive and much of the material could not be covered in a semester course. It still is a good reference book. The book has 1196 pages, has no pictures, and has comparatively few Tables and graphs. Tanner also published a laboratory workbook on food microbiology.

W. C. Frazier's book (*Food Microbiology* published by McGraw-Hill Book Co., New York), since its publication in 1958 (2nd edition, 1967), has replaced Tanner's *Microbiology of Foods* as a textbook. This book has 537 pages and is organized into 6 parts and 30 chapters. The six parts are: Microorganisms Important in Food Microbiology; Principles of Food Preservation and Spoilage; Contamination, Preservation and Spoilage of Different Kinds of Foods; Foods and Enzymes Produced by Microorganisms; Foods in Relation to Disease and Food Sanitation; Control and Inspection. This book is illustrated with pictures, Tables and figures and is considerably shorter than Tanner's book but longer than one by Jay and another by Nickerson and Sinskey.

James Jay's book (*Modern Food Microbiology*), published in 1970 by Van Nostrand Reinhold Co., New York, contains 328 pages and is illustrated by pictures, Tables and figures. It is divided into 20 chapters. The chapter titles are: History of Microorganisms in Foods; The Role and Significance of Microorganisms in Nature and in Foods; Intrinsic and Extrinsic Parameters of Foods that Affect Their Microbiology; The Incidence and Types of Microorganisms in Foods; Food Spoilage: Spoilage of Market Fruits and Vegetables; Food Spoilage: Spoilage of Fresh and Cured Meats, Poultry and Sea Foods; Spoilage of Miscellaneous Foods; Food Preservation by the Use of Chemicals; Food Preservation by Use of Radiation; Food Preservation by Use of Low Temperatures; Food Preservation by Use of High Temperatures; Preservation of Foods by Drying; Indices of Food Sanitary Quality, and Microbiological Standards; Food Poisoning Caused by Gram-positive Cocci; Food Poisoning Caused by Gram-positive Sporeforming Bacteria; Food Poisoning Caused by Gram-negative Bacteria; Biological Hazards in Foods Other than Food Poisoning Bacteria; Characteristics and Growth of Psychrophilic Microorganisms; Characteristics and Growth of Thermophilic Microorganisms; and Nature of Radiation Resistance in Microorganisms.

The textbook of Nickerson and Sinskey published in 1972 is a mixture of microbiology and food processing as the name implies (*Microbiology of Foods and Food Processing* published by North-Holland Publishing Co., New York). It contains 307 pages illustrated by pictures, Tables and figures. The 12 chapter titles are as follows: Methods Used for the Microbiological Examination of Foods; The Destruction of Bacteria by Heat; Microbiology of Dried Foods; The Growth of Microorganisms in Foods at Refrigerator Temperatures above Freezing; The Preservation of Foods by Freezing; Chemical Preservation of Foods; Microbiology of Flesh-type Foods and Eggs; Fruits and Vegetables; Microbiology of Dairy Products; Food Infections; Food Intoxications; and Sanitation in Food-manufacturing Plants and in Food-preparation and Serving Establishments.

Specific page numbers of pertinent data found in each of the above books are given in Table 1.5.

TABLE 1.5. SUBJECTS OF IMPORTANCE IN OTHER TEXTBOOKS WHICH ARE DISCUSSED IN THE FOLLOWING CHAPTERS[1]

Subject	Frazier (1967) (pp. nos.)	Jay (1970) (pp. nos.)	Nickerson and Sinskey (1972) (pp. nos.)
Bacteria	36–62	12–18	
Yeast	25–35	22–24	
Molds	2–24	18–22	
Factors influencing growth of spoilage microorganisms	160–174	26–39	
Fruits and vegetables	201–248		169–174
Red meats	252–280	68–80	137–150
		41–46	
Poultry meats	310–316	46	150–152
		80–81	
Fish and sea foods	283–294	46–47	152–161
		81–86	
Eggs and egg products	296–308	89–91	161–166
		241–242	
Dehydrated fruits, vegetables and cereals	122–129	91	
Canned foods	357–366	98–102	26–69
	82–107		
Chemically preserved	131–144	106	119–135
foods and UV light	149–151	122	
Fermented foods	380–414	50, 92–93	
Milk and milk products	318–348	51, 242, 188	176–195
		164, 3, 5, 17	
Tests for organisms in foods	492–500		1–24
Refrigerated foods	115–120	164–165	
		143, 137–149	
Food poisoning and infection	434–472	194–253	197–266

[1] In addition to the list of references at the end of this chapter for more complete information and further study, attention is called to Weiser *et al.* (1971); and for a compendium of methods for the microbiological examination of foods see Speck (1976).

Examples of Some Journals which Publish Articles Dealing with Food Microbiological Subjects

Applied Microbiology published by American Society for Microbiology, 19131 St, NW, Washington, D.C. 20006.

Journal of Applied Bacteriology published by Academic Press, 111 Fifth Ave., New York, N.Y.

Journal of Food Science published by Institute of Food Technologists, 221 North LaSalle St., Chicago, Illinois 60601.

Food Technology, published by Institute of Food Technologists, 221 North LaSalle St., Chicago, Illinois 60601.

American Journal of Public Health published by American Public Health Association, 1015 18th St., NW, Washington, D.C. 20036.

Journal of Dairy Science published by American Dairy Science Association, 113 North Neil St., Champaign, Illinois 61820.

Fisheries and Environment, Canada Fisheries and Marine Service,

Scientific Information and Publications Branch, Ottawa, K1ADE6, Canada.

Australian Journal of Applied Science, Commonwealth Scientific and Industrial Research Organization, 314 Albert St., East Melbourne, C. 2, Victoria, Australia.

Poultry Science published by The Poultry Science Association, c/o C. B. Ryan, Sec.-Treas., Texas A&M University, College Station, Texas 77843.

Journal of Milk and Food Technology published by International Association of Milk, Food and Environmental Sanitarians, P.O. Box 437, Blue Ridge Road, Shelbyville, Indiana 46176.

REFERENCES

BITTING, A. W. 1937. Canning. *In* Food Technology. McGraw-Hill Book Co., New York.

BUCHANAN, B. F. 1972. The Mission of food science programs: an industry view. Food Technol. *26,* 76–81.

DESROSIER, N. W. 1977. The Technology of Food Preservation, 4th Edition. AVI Publishing Co., Westport, Conn.

FRAZIER, W. C. 1967. Food Microbiology, 2nd Edition. McGraw-Hill Book Co., New York.

JAY, J. M. 1970. Modern Food Microbiology. Van Nostrand Reinhold Co., New York.

MILNER, M., SCRIMSHAW, M.D. AND WANG, D. I. C. 1978. Protein Resources and Technology. AVI Publishing Co., Westport, Conn.

NICKERSON, J. T. and SINSKEY, A. J. 1972. Microbiology of Foods and Food Processing. American Elsevier Publishing Co., New York.

ROBERTS, W. M. *et al.* 1969. Recommendations on subject matter outline for food oriented courses. J. Food Technol. *23,* 307–311.

SCRIMSHAW, N. S. and HEGSTED, D. M. 1972. Research in Nutrition and Food Science. *In* Report of the Committee on Research Advisory to the U.S. Department of Agriculture. National Research Council. Distributed by National Technical Information Service, U.S. Dept. of Commerce, Springfield, Virginia.

SPECK, M. L. (Editor). 1976. Compendium of Methods for the Microbiological Examination of Foods. American Public Health Assoc., Washington, D.C.

TANNER, F. W. 1944. The Microbiology of Foods. Garrard Press, Champaign, Illinois.

WEISER, H. H., MOUNTNEY, G. J. and GOULD, W. A. 1971. Practical Food Microbiology and Technology, 2nd Edition. AVI Publishing Co., Westport, Conn.

2

Bacteria

OCCURRENCE AND IMPORTANCE

Food Spoilage Bacteria

Dominant Flora Concept.—According to Mossel and Ingram (1955), there are a definite number of genera that cause foods to spoil under standard conditions of storage. This was a principle formulated by Beijerinck, a famous bacteriologist. The concept of a dominant spoilage microflora will be repeated through this book. Classes of foods and the dominant flora under standard conditions of storage are listed in Table 2.1. Note the occurrence of the genera *Micrococcus, Pseudomonas,* and *Flavobacterium.* Other bacteria which are involved in the spoilage of foods are given in Table 2.2.

Food Poisoning and Infection.—Food poisoning bacteria include *Clostridium botulinum* and *Staphylococcus aureus,* whereas *Salmonella* and *Shigella* are considered infectious. Experts are not sure whether *Clostridium perfringens* causes a food poisoning and/or an infection. More will be said about these in the chapter on food poisoning.

Plant Pathogens.—These are bacteria which cause plant diseases. These may occur in the field but also affect the fruit or vegetable during their marketing. A good example of such a bacterium is *Erwinia carotovora* which causes soft rot in vegetables.

ISOLATION PROCEDURES

The isolation of bacteria may be achieved by streaking the liquid from a broth culture on a solid growth medium (like nutrient agar) until isolated

18

TABLE 2.1.   DOMINANT BACTERIA SPOILING MOST IMPORTANT FOODS

Class of Food Products	Genera Dominating When Spoilage Occurs During Standard Conditions of Storage
Milk and milk products	*Streptococcus, Lactobacillus, Microbacterium,* Gram-negative rods,[1] *Bacillus*
Fresh meat	Gram-negative rods,[1] *Micrococcus*
Poultry	Gram-negative rods,[1] *Micrococcus*
Sausage, bacon, ham	*Micrococcus, Lactobacillus, Streptococcus*
Fish, shrimp	Gram-negative rods,[1] *Micrococcus*
Shellfish	Gram-negative rods,[1] *Micrococcus*
Vegetables	Gram-negative rods,[1] *Lactobacillus, Bacillus*
Bread	*Bacillus*

Source: Mossel and Ingram (1955).
[1] Gram-negative rods mean *Pseudomonas* and *Flavobacterium* with the exclusion of the coli-enterobacter group.

colonies are obtained. These isolated colonies are considered to be "pure" and may be cultured as "pure" and then identified.

Another way to obtain isolated colonies is to use the pour plate method. Dilutions of the original material are made, aliquots are added to Petri plates, and a nutrient agar is added. Isolated colonies are handled as above. Still another way is to make appropriate dilutions and add an aliquot to the surface of agar on a previously poured plate. Isolated colonies are cultivated as before.

The procedure just described applies to both aerobes and anaerobes. Of course, anaerobes require an oxygen-free environment such as can be developed in an anaerobic incubator.

MEDIA FOR ISOLATION, CULTIVATION AND IDENTIFICATION

Many times specific media are needed to isolate, cultivate and identify bacteria. For example, many bacteria do not grow well on nutrient agar so that medium would not be the one of choice. Media listed in Table 2.3 are recommended to isolate the dominant flora which are given in Table 2.1. Specific media for the isolation, culture and identification of coliforms, *Salmonella* and *Shigella* are listed in Table 2.4. See the manuals of Difco and Baltimore Biological Laboratories for additional substrates.

TABLE 2.2.   SPOILAGE OF FOODS BY BACTERIA

Bacterium	Occurrence and/or Importance
Pseudomonas fluorescens	Soil, water, eggs, cured meats, fish, milk
Alcaligenes faecalis	Milk, rotting eggs, water decomposing organic matter
Brucella spp.	Causes brucellosis in man
Escherichia coli	Indicator of fecal pollution; some serotypes cause infections, enteric diseases in human infants
Salmonella spp.	Food infections in man; diseases in animals
Shigella	Intestinal tract of man and animals; causes dysentery
Serratia marcescens	Found in fecal matter, soil and decomposing animal matter
Erwinia carotovora	Causes rotting of plant matter
Flavobacterium spp.	Soil and water; spoils vegetables
Micrococcus roseus	Found in dust, water and foods containing salt; produces a red sediment in milk
Micrococcus varians	Milk, dairy products, dust, soil, meat
Staphylococcus aureus	Infections, boils, food poisoning
Streptococcus agalactiae	Milk, cows with mastitis
S. faecalis	Feces of humans and warm-blooded animals; fecal indicator
S. lactis	Common in milk and dairy products; causes milk to sour
S. cremoris	Milk and milk products; used in cultured buttermilk
S. thermophilus	Milk and milk products; causes milk to sour
Leuconostoc mesenteroides	Occurs in normal fermentation of plant materials; found in slimy sugar solutions, milk and dairy products
Pediococcus cerevisiae	Produces undesirable flavor in kraut; spoils beer
Bacillus subtilis	Occurs in various kinds of plant and animal materials if oxygen is available and found in soil; causes ropiness in bread; spoilage of home-canned green beans, tomatoes
B. cereus	Found in soil; causes food poisoning; causes acid proteolysis in milk; spoilage of home-canned tomatoes and green beans; spoilage of canned evaporated milk
B. licheniformis	Found in soil; causes ropiness in bread; spoilage of home-canned green beans and tomatoes
B. megaterium	Found in soil; causes spoilage of canned evaporated milk
B. coagulans	Spoils canned evaporated milk; flat-sour spoilage of tomatoes and tomato juice
B. stearothermophilus	Found in soil; flat-sour spoilage of low-acid, canned vegetables
Clostridium butyricum	Spoils canned tomatoes and pineapple; spoils green olives

TABLE 2.2. (*Continued*)

Bacterium	Occurrence and/or Importance
C. thermosaccharolyticum	Spoils canned foods
C. pasteurianum	Spoils canned tomatoes, pears and pineapple
C. sporogenes	Spoils medium- and low-acid canned foods
C. botulinum	Spoils medium- and low-acid foods and causes botulism
Desulfotomaculum nigrificans	Causes sulfide-stinkers in low-acid foods

Source: Buchanan and Gibbons (1974); Frazier (1967).

TABLE 2.3. SELECTED MEDIA FOR ISOLATION, CULTIVATION AND IDENTIFICATION OF DOMINANT SPOILAGE FLORA[1]

Bacteria	Isolation	Cultivation	Identification
Streptococcus	Eugonbroth Eugonagar TSA[2] or TSB[3]	Eugonagar Eugonbroth	Thiogel medium
Lactobacillus	Tomato juice agar Eugonagar	Eugonagar agar	CTA medium Indole-nitrite medium Litmus, purple and Ulrich milks
Pseudomonas Flavobacterium Micrococcus	Nutrient agar[4]	Nutrient agar[4]	Nitrate broth[4] Litmus milk[4] Nutrient agar[4] Indole medium[4]
Bacillus	Nutrient agar[4]	Nutrient agar[4]	Thiogel medium[4] Litmus milk[4]
Clostridium	Anaerobic agar Thioglycollate medium	Anaerobic agar	Anaerobic agar w/o dextrose Iron milk Indole-nitrite medium
Microbacterium	TSA[2] Nutrient agar	TSA[2] Nutrient agar	

Source: BBL (1956).
[1] See Table 2.1.
[2] Trypticase soy agar.
[3] Trypticase soy broth.
[4] Not listed by BBL (1956).

CLASSIFICATION OF BACTERIA

By Physiological Reactions

Bacteria in food microbiology can be grouped by their physiological characteristics. For example, the type of acid produced may be used as a category. Many of the food spoilage genera have lipolytic, pectolytic or proteolytic enzymes. Different bacteria spoil canned foods, for example, than spoil refrigerated foods. Tolerance to salt and sugar also varies. A grouping of bacteria based on physiological reactions is given in Table 2.5.

TABLE 2.4.   SELECTED MEDIA FOR ISOLATION, CULTIVATION, IDENTIFICATION OF COLIFORMS, *SALMONELLA* AND *SHIGELLA*

Bacteria	Isolation	Media Cultivation	Identification
Coliforms	Brilliant-green-bile media[1]	TSA[2]	MR-VP medium[1]
	Lactose broth[1]	TSB[3]	
Salmonella	Desoxycholate-citrate agar[1]	TSA[2]	Triple-sugar-iron agar[4]
			Urease test medium[1]
Shigella	Desoxycholate-citrate agar[1]	TSA[2]	Triple-sugar-iron agar[4]
			Urease test medium[1]

[1] BBL (1956).
[2] Trypticase soy agar (BBL 1956).
[3] Trypticase soy broth (BBL 1956).
[4] Not listed in BBL (1956).

The food microbiologist may want to know how many proteolytic bacteria there are in foods and may use milk and/or gelatin plating media to enumerate these organisms. The same is true for lipolytic bacteria. Other studies may involve apparatus similar to that shown in Fig. 2.1.

According to Bergey's Manual

Classification of bacteria according to the Eighth Edition of *Bergey's Manual of Determinative Bacteriology* (Buchanan and Gibbons, 1974), highlights the following data.

Key to Families

Gram-negative rods	
Aerobic	Pseudomonadaceae
Facultatively anaerobic	Enterobacteriaceae
Gram-positive cocci	
Cells in regular or irregular clusters or packets	Micrococcaceae
Cells in pairs, chains or tetrads	Streptococcaceae
Gram-positive rods producing endospores	Bacillaceae
Gram-positive rods not producing endospores	Lactobacillaceae

TABLE 2.5.   CLASSIFICATION OF BACTERIA BASED UPON PHYSIOLOGICAL
REACTIONS

Physiological Character	Examples in the Following Genera
Produce lactic acid	Streptococcus
	Leuconostoc
	Pediococcus
	Lactobacillus
Produce acetic acid	Acetobacter
Produce butyric acid	Clostridium
Produce propionic acid	Propionibacterium
Produce proteolytic enzymes	Bacillus
	Pseudomonas
	Clostridium
	Streptococcus
	Micrococcus
	Proteus
Produce lipolytic enzymes	Pseudomonas
	Alcaligenes
	Micrococcus
Produce pectolytic enzymes	Erwinia
	Bacillus
	Clostridium
Grow at 55°C (thermophilic)	Bacillus
	Clostridium
Grow at temperatures near 0°C (psychrophilic)	Pseudomonas
	Alcaligenes
	Flavobacterium
Produce gas	Leuconostoc
	Lactobacillus
	Propionibacterium
	Escherichia
	Enterobacter
	Proteus
	Bacillus
	Clostridium
Pigmented bacteria	Flavobacterium
	Serratia
	Micrococcus
Halophilic	Pseudomonas
	Micrococcus
	Pediococcus
Sugar-tolerant	Leuconostoc

Source: Frazier (1967).

GRAM-NEGATIVE AEROBIC RODS AND COCCI

1.   Family Pseudomonadaceae (Buchanan and Gibbons 1974).

These bacteria are motile rods by polar flagella. They have a respiratory metabolism, not fermentative. The rods are strict aerobes which are catalase-positive and usually oxidase-positive. Their G + C (guanine + cystosine) content is 50–70 moles %.

Strains of Pseudomonas are found in soil and fresh water as well as in marine environments. These bacteria are a part of the dominant spoilage flora (see Table 2.1) in several food groups.

Courtesy of B. Tweedy, University of Missouri-Columbia

FIG. 2.1. RADIORESPIROMETER APPARATUS
USED TO STUDY GLUCOSE METABOLISM IN
FOOD SPOILAGE BACTERIA

a. Genus *Pseudomonas.* Many of the species of *Pseudomonas* do not require any growth factors (can grow in mineral media with a single organic compound as sole source of carbon). Most strains grow at 4°C or below. These bacteria cause spoilage of refrigerated foods.

b. Genera of uncertain affiliation.

i. *Alcaligenes.* These rods are motile with peritrichous flagella. These strictly aerobic saprophytic bacteria are not actively proteolytic. They occur in the intestinal tract of vertebrates, in dairy products, in rotten eggs and fresh water.

ii. *Acetobacter.* The rods also occur in involuted forms (spherical, swollen, club-shaped, curved, branched or filamentous) which are motile by peritrichous flagella. Some are non-motile. They oxidize ethanol under strict aerobic conditions to form acetic acid.

iii. *Brucella.* These coccobacilli are non-motile and cause undulant fever in man, goats, and swine.

GRAM-NEGATIVE FACULTATIVELY ANAEROBIC RODS

1. Family Enterobacteriaceae (Buchanan and Gibbons 1974).

The Enterobacteriaceae may be motile by peritrichous flagella or non-motile; their metabolism is either oxidative or fermentative. Acid is produced by the fermentation of glucose; catalase is positive and nitrates are reduced to nitrites.

a. Genus *Escherichia*. *Escherichia coli* is the only species of the genus. This bacterium is present in the intestines of warm-blooded animals. Some strains are opportunistic pathogens. This species is used as an indicator bacterium for fecal pollutions.

b. Genus *Salmonella*. There are many serological types. Strains are given names of towns, regions, etc., where the strain was isolated. These bacteria cause disease and food infections.

c. Genus *Shigella*. These bacteria cause dysentery which may be spread through food or water.

d. Genus *Serratia*. Strains of *Serratia* may be pink, red or magenta. These bacteria may produce colors on foods as they grow.

e. Genus *Proteus*. These microbes may be present in foods due to fecal contamination. Strains are proteolytic and may be urease positive.

f. Genus *Erwinia*. These bacteria are associated with plants. The group which food microbiologists are interested in is the carotovora group which is pectolytic. The major species is *E. carotovora* which causes a wide range of spoilage in vegetables, especially in storage and marketing of the fresh material.

g. Genus *Enterobacter*. These motile rods belong to the coliform group which is associated with plant materials rather than with fecal matter. They can use citrate and acetate as the sole carbon source.

h. Genera of uncertain affiliation.

ii. *Flavobacterium*. Colonies of these bacteria are yellow, red, brown. The hue varies with the growth medium and temperature. These pigments are not soluble in media. Flavobacteria are oxidative rather than fermentative. They are common in water and soil and contaminate vegetables, milk and milk products, fresh meat, poultry, fish, shrimp and shellfish (see Table 2.1). Bacteria are part of the dominant spoilage flora on these foods.

GRAM-POSITIVE COCCI

1. Family Micrococcaceae (Buchanan and Gibbons 1974).

These Gram-positive cocci divide in more than one plane to form regular or irregular clusters or packets. These microorganisms form acid but no gas from glucose. They have a tolerance for sodium chloride. All grow in

5% and some in as much as 10–15% sodium chloride. These bacteria are catalase positive, aerobic to facultatively anaerobic.

a. Genus *Micrococcus*. Micrococci have strictly respiratory metabolism. *M. luteus* is common in soil, dust, water and on the skin of animals and man. *M. roseus* produces pink or red colonies. These bacteria occur in dust, water and salt-containing foods. *M. varians* forms yellow colonies. They are common in milk, animal carcasses, dust and soils. The genus *Micrococcus* is part of the dominant spoilage flora in several food groups (Table 2.1).

b. Genus *Staphylococcus*. These cocci produce cell masses that look like grapes, hence the name meaning the grape-like coccus. Their metabolism is both respiratory and fermentative. These facultative anaerobes produce catalase and can ferment a wide range of carbohydrates. The important species is *S. aureus* which causes food poisoning. Its colonies are white, yellow or orange. Coagulase is produced by most strains. *S. aureus* is salt tolerant. All strains are potential pathogens.

2. Family Streptococcaceae (Buchanan and Gibbons 1974).

Bacteria in this family are spherical or ovoid facultative anaerobes with generally complex nutritional requirements. They do not form spores and are rarely motile.

Streptococcaceae are found on dairy utensils, manure, saliva, intestinal tract of man and animals, feeds and certain plants.

Bacteria in this family are important because of their role in the fermentation of sauerkraut, pickles, cheese, fermented milk, cultured butter and cream. They are part of the dominant spoilage flora in fluid milk and milk products (Table 2.1) as well as in fruit juice concentrates, ketchup, cream and canned fruit.

a. Genus *Streptococcus*. *S. pyogenes* is a human pathogen which produces beta-hemolysis on blood agar plates. *S. agalactiae* causes mastitis in cows and produces alpha-hemolysis when grown on blood agar. *S. thermophilus* is important in the making of Swiss cheese and yogurt. *S. bovis* occurs in human feces and in the alimentary tract of ruminants. *S. lactis* and *S. cremoris* are used in the manufacture of buttermilk and cheese. *S. faecalis* is found in the intestinal tract of man and animals. This bacterium has been used as a fecal indicator.

b. Genus *Leuconostoc*. The species *L. mesenteroides* causes slime in sugar refineries; this bacterium is important in the fermentation of cabbage, cucumber and olives. *L. mesenteroides* is also found on fruits and in dairy products. *L. cremoris* is used as a starter culture in milk and dairy products.

c. Genus *Pediococcus*. *P. cerevisiae* spoils beer by producing diacetyl. This bacterium is highly tolerant of the antiseptics in hops. *P. acidilactici* grows in unhopped beer but not in hopped beer. *P. pentosaceus* occurs in fermenting cereal mashes, cabbage and cucumbers. *P. halophilus* grows in soy mash, soybean cheese, and anchovies.

ENDOSPORE-FORMING RODS

1. Family Bacillaceae (Buchanan and Gibbons 1974).

This family is characterized by the formation of endospores. Young cells of *Bacillus megaterium* contain vacuoles as illustrated in Fig. 2.2. The rods are motile by lateral or peritrichous flagella; these are shown in Fig. 2.3 and 2.4. Included among the Bacillaceae are aerobes and facultative anaerobes, and strict anaerobes.

a. Genus *Bacillus*. There is a great diversity in properties within the genus (growth requirements, temperature of growth and salt tolerance). Species of *Bacillus* are aerobic and/or facultative anaerobes.

This genus is part of the dominant spoilage flora in milk and milk products and vegetables; see Table 2.1 for specific foods.

b. Genus *Clostridium*. Species of this genus are anaerobic. As shown in Table 2.2, clostridia can cause spoilage of canned foods and meat products.

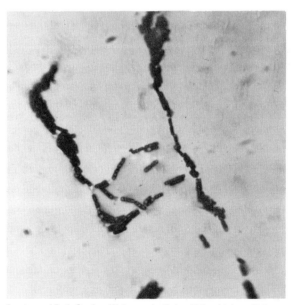

Courtesy of Ruth Gordon, Waksman Institute of Microbiology, Rutgers

FIG. 2.2.   *BACILLUS   MEGATERIUM*   ATCC   14581   ON GLUCOSE AGAR, 1 DAY
Note the vacuoles in the cells. × 1575.

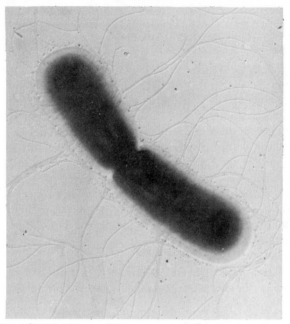

FIG. 2.3.   *BACILLUS MEGATERIUM* ATCC 14581 PERITRI-
CHOUSLY AND ABUNDANTLY FLAGELLATED. × 16,000

c. Genus *Desulfotomaculum*. These Gram-negative rods reduce sulfates and sulfites to H_2S. Sulfur is required for growth of these strict anaerobes.

D. nigrificans causes sulfide-stinker spoilage of canned foods.

GRAM-POSITIVE ASPOROGENOUS ROD-SHAPED BACTERIA

Lactobacilli are rarely motile and are facultative to anaerobic. They have complex organic nutritional requirements. They are highly fermentative with the production of lactic acid. These bacteria are divided into two groups—homofermentative and heterofermentative. The homofermentative group produces mostly lactic acid from glucose. Examples of this group are: *L. delbrueckii, L. lactis, L. bulgaricus, L. plantarum.*

Heterofermentative lactobacilli are: *L. hilgardii, L. fructivorans, L. desidiosus,* and *L. ruminis*.

As indicated in Table 2.1, members of the genus *Lactobacillus* are part

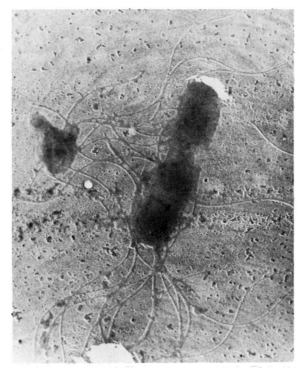

Courtesy of University of Missouri-Columbia

FIG. 2.4. *BACILLUS STEAROTHERMOPHILUS* × 20,000

of the dominant spoilage flora in milk and milk products, sausage, bacon, ham, and vegetables.

PRESERVATION OF BACTERIAL CULTURES

Bacterial cultures can be maintained on various agars (Table 2.3). The more fastidious cultures (nutritionally) will have to be transferred every month or two, whereas the spore-forming rods (*Bacillus* and *Clostridium*) can be kept for years if the media do not dry out.

Bacteria may be freeze-dried in milk and kept for long periods of time. Here again, different species react differently to this type of treatment; some are more sensitive than others.

Bacteria may also be preserved on agar slants over which sterile mineral oil has been placed.

GLOSSARY

Aerobic, aerobes—Bacteria that live in the presence of molecular oxygen.

Anaerobic, anaerobes—Bacteria that live in the absence of molecular oxygen.

Asporogenous—Bacteria which do not form endospores.

Catalase—An enzyme that breaks down hydrogen peroxide to oxygen and water. A catalase-positive bacterium has this enzyme.

Dominant spoilage flora—These are the genera that dominate and are responsible for the spoilage of a food.

Endospores—Endospores are intracellular bodies, resistant to heat, dryness and chemicals.

Facultative anaerobes or facultative—These are bacteria that can live either in the presence or absence of molecular oxygen.

Fermentative metabolism—A metabolism in which useful by-products such as acid and/or alcohol are produced. This is contrasted to respiratory metabolism in which the end products are carbon dioxide and water.

Flagella—Plural for flagellum. A whip-like extension of a bacterium functioning in cell motility.

Food infection—An infection obtained by eating food containing bacteria such as *Salmonella*. See food poisoning.

Food poisoning—Bacteria such as *Clostridium botulinum* or *Staphylococcus aureus* grow and produce a toxin in the food. When the food is eaten, the person becomes ill. The individual does not need to eat the live bacteria as in food infections.

G+C content—This stands for the guanine and cytosine content of the DNA in the cell.

Halophilic—Salt-loving.

Hemolysis—To lyse or to dissolve red blood cells.

Heterofermentative—The bacteria produce many by-products of fermentation.

Homofermentative—The bacteria produce mainly (85% or greater) one product of fermentation, such as lactic acid.

Lipolytic enzymes—The breakdown of fats or lipids by enzymes.

Oxidase-positive—The bacteria produce oxidase enzymes.

Pectolytic enzymes—Enzymes which hydrolyze pectin.

Peritrichous flagella—Flagella located on all sides of the cell.

Proteolytic enzymes—Enzymes that break down proteins.

Psychrophilic bacteria—Bacteria that love to grow at cold temperatures.

Respiratory metabolism—Oxidative processes that finally convert carbon compounds into carbon dioxide and water.

Serological types—Bacteria classified according to reaction in blood serum.

Soft rot—The rot produced by the growth of pectolytic bacteria in plant tissue. The loss of pectin makes the plant very soft.

Sporogenous—Produces spores.

Thermophiles—Bacteria which love to grow at high temperatures.

Urease positive—The bacteria produce enzymes to break down urea.

REVIEW QUESTIONS

1. What is the "dominant flora" concept?
2. List food groups and their dominant spoilage flora.
3. Differentiate between food poisoning and food infections.
4. Make a list of food spoilage bacteria and their occurrence.
5. Make a list of media needed to isolate, cultivate and identify the dominant food spoilage bacteria.
6. List bacteria according to physiological characteristics.
7. Give a simple key to separate the following families: Pseudomonadaceae, Enterobacteriaceae, Micrococcaceae, Streptococcaceae, Bacillaceae and Lactobacillaceae.
8. Give the significance of each of the following bacteria as they relate to foods:
   a. *Escherichia coli*
   b. *Salmonella*
   c. *Shigella*
   d. *Erwinia carotovora*
   e. *Proteus*
   f. *Staphylococcus aureus*
   g. *Streptococcus agalactiae*
   h. *Leuconostoc mesenteroides*
   i. *Pedicoccus cerevisiae*
   j. *Bacillus coagulans*
   k. *Clostridium botulinum*

SUGGESTED LABORATORY EXERCISES

Take representatives from each family, using the most important species, and study these in the laboratory to become thoroughly familiar with the bacteria and the media used to grow and identify the bacteria.

By midterm, each student should be tested on the bacteria, the media, and the techniques needed in isolation and identification of these microorganisms. The final part of the semester may be devoted to isolating and identifying mixtures of selected bacteria, molds and yeasts from milk.

REFERENCES

BBL. 1956. BBL Products, Culture Media, Materials and Apparatus for the Microbiological Laboratory. Baltimore Biological Laboratories, Inc., Baltimore.

BUCHANAN, R. E. and GIBBONS, N. E. (EDITORS). 1974. Bergey's Manual of Determinative Bacteriology, 8th Edition. The Williams and Wilkins Co., Baltimore.

DIFCO LABORATORIES. 1953. Difco Manual of Dehydrated Culture Media and Reagents for Microbiological and Chemical Laboratory Procedures, 9th Edition. Difco Laboratories, Detroit.

FRAZIER, W. C. 1967. Food Microbiology. McGraw-Hill Book Co., New York.

MOSSEL, D. A. A. and INGRAM, M. 1955. The physiology of the microbial spoilage of foods. J. Appl. Bacteriol. *18*, 232–268.

3

Yeasts

OCCURRENCE AND IMPORTANCE

Ecology of Nonhuman and Nonanimal Pathogens

The yeasts are widely distributed in nature. They occur on leaves, flowers, sweet fruits (strawberries, cherries, plums, gooseberries, and grapes), grains (wheat, barley, oats, corn), fleshy fungi (mushrooms), exudates of trees (slime flux of oaks, elms, horse chestnuts, and limes; exudates of pine trees) and in the upper layers of soil in orchards and vineyards. Insects, especially fruit flies, spread yeasts which are essential to their nutrition.

Rhodotorula glutinis and *Hansenula anomala* have passed through the human GI tract without being damaged by digestive juices (Lund 1958) and appear in the feces.

Nematospora coryli causes diseases in hazelnuts, tomatoes, coffee beans, pecan nuts, lima beans, soybeans, citrus fruits and pomegranates. Citrus fruits are rotted by *Nematospora*; *Kloeckera apiculata* also causes a soft rot of strawberries (Ingram 1958).

A summary of yeasts which spoil particular foodstuffs is given in Table 3.1. As illustrated by data in this Table, yeasts are not limited to fruits and fruit products but also occur in cereals, meat and dairy products.

Ecology of Human and Animal Pathogens

These yeasts are out of the scope of this course and are mentioned here only to show their existence and to complete our outline.

Industrial Importance of Yeasts

Because of their ability to ferment carbohydrates, yeasts are used industrially to:

1. Produce beer and other alcoholic beverages.
2. Make wines.

3. Make bakery products.
4. Produce protein, fat and vitamins.
5. Produce enzymes.
6. Produce food for man and animals.

ISOLATION PROCEDURES

In the study of yeasts in nature, it is necessary to use certain inhibitors of bacteria and molds to prevent overgrowth of the yeasts present. The following methods have been used.

Place material from which yeasts are to be isolated into a suitable medium such as malt broth and incubate at 25°C. However, rapid growers may outgrow slow growers with this technique. If material is shaken, molds in the sample will grow in balls, making easier the isolation of the yeasts by streaking.

If the enrichment and/or plating medium has a pH of 3.5 to 3.8, competing bacteria can be eliminated.

A broad-spectrum antibiotic can also be used in the media to inhibit the bacteria.

TABLE 3.1. SPOILAGE OF FOODS BY YEASTS

Foods	Food Spoilage Yeasts
Cereal	
Bin-stored rice	*Hansenula anomala*
	Candida tropicalis
	Candida pseudotropicalis
	Pichia farinosa
Fruits and juices	
Citrus fruits	*Nematospora coryli*
Strawberries	*Kloeckera apiculata*
Apple juice	*Saccharomyces* sp.
Citrus juice	*Saccharomyces cerevisiae*
Grape juice	*Saccharomyces cerevisiae*
Dehydrated, concentrated foods and fermented foods	
Dried fruits, juice concentrate, sugar, sugar products and honey	*Saccharomyces rouxii*
Sweet pickles	*Saccharomyces bisporus* var. *mellis*
Brined cucumbers	*Hansenula subpelliculosa*
Meat and meat brines	
Sausage	*Debaryomyces guilliermondii* var. *nova-zeelandicus*
Chlortetracycline-treated meat	*Candida lipolytica*
Meat-curing brines	*Debaryomyces membranaefaciens* var. *hollandicus*
Dairy products	
Milk, sweetened condensed milk	*Torulopsis*
Butter	*Rhodotorula*
Cheese	*Kluyveromyces fragilis*

Source: Ingram (1958).

The following materials can be used to prevent the growth of molds.

1. 0.01% diphenyl
2. 0.25–0.35% sodium propionate
3. 1% oxgall
4. 0.003% rose bengal

Osmophilic yeasts may be isolated on agar containing 30% glucose. If source material is a syrup, use 50% glucose as the diluent.

MEDIA FOR CULTIVATION

Difco Laboratories of Detroit manufacture potato dextrose, malt and wort agars for general cultivation of yeasts. All three of these media can also be purchased from Baltimore Biological Laboratory, Baltimore.

Specific substrates are required for determining the species of yeasts. Van der Walt (1971A) listed various media, as shown in Table 3.2. The formulas are given in Table 3.3.

For proper nutrition for the yeasts, media should contain a carbon source, an organic or inorganic nitrogen source, various minerals and a mixture of

TABLE 3.2.   MEDIA FOR STUDYING YEASTS

Factor	Medium
Isolation and culture maintenance	Yeast extract-malt extract broth; yeast extract-malt extract agar and malt agar.
Morphological characteristics (vegetative reproduction and characteristics of vegetative cells)	Malt extract or glucose-yeast extract-peptone water; malt agar; glucose-yeast extract-peptone agar; Bacto-yeast morphology agar.
Formation of pseudomycelium and true mycelium	Corn meal agar and potato-glucose agar.
Growth on liquid media	Malt extract broth or glucose-yeast extract-peptone water.
Growth on solid media	Malt extract agar; 2% glucose-yeast extract-peptone agar.
Production of ascospores	Yeast-malt extract (YM) agar.
Utilization of nitrogen compounds	Yeast-carbon base medium.
Utilization of carbon compounds	Yeast-nitrogen base medium.
Utilization of vitamins	Vitamin-free yeast base.

Source: Van der Walt (1971A).

TABLE 3.3.   INGREDIENTS OF MEDIA LISTED IN TABLE 3.2

Medium	pH	Ingredients	g/liter
Yeast extract-malt extract broth	5–6	Yeast extract	3
		Malt extract	3
		Peptone	5
		Glucose	10
		Agar added	(2% agar)
Malt extract agar[1]	4.7	Malt extract	15
		Agar	20
Glucose-yeast extract-	not	Glucose	20
peptone water	adjusted	Peptone	10
		Yeast extract	5
		Agar added	20
Bacto-yeast morphology agar	4.5	Mixture[2]	35
Cornmeal agar[1]	6.0	Infusion from cornmeal	50
		Agar	15
Potato dextrose agar[1]	5.6	Infusion from potato	20
		Glucose (dextrose)	20
		Agar	15

Source: Van der Walt (1971A).
[1] Difco (ingredients as listed in Difco Manual, 1953).
[2] See Difco Manual (1953) page 251 for exact ingredients. Contains N, C sources plus amino acids, vitamins, minerals (trace), salts and agar.

vitamins. The optimum pH is usually 4.5 to 6.5, but yeast can grow over a range of 3 to 8.

The yeast-nitrogen base (manufactured by Difco Laboratories and used to test carbohydrate utilization contains the following ingredients: trace elements, nine vitamins, trace of amino acids, potassium magnesium sulfate, sodium chloride and calcium chloride. The nitrogen source is ammonium sulfate.

The yeast-carbon base medium is used for testing growth on various nitrogen sources. This medium is identical with the nitrogen base except the carbon source (glucose) is supplied and the ammonium sulfate is deleted. Possible nitrogen sources include amino acids, purines and pyrimidine bases, amines, urea, nitrate, nitrite and ammonium ion in the form of a salt.

The base media are poured into Petri plates along with the yeast being tested. Several compounds may be spotted on the same plate. If the yeast can assimilate the sugar or nitrogen source, growth will occur where the material was added to the agar.

Special media are needed sometimes to induce sporulation; carrot plugs, raisin agar, prune agar, and orange juice agar have all been used. Many yeasts sporulate easily and readily on yeast extract-malt extract (YM) agar. Gorodkuowa agar (0.1% glucose; 1% peptone, 0.5% sodium chloride and 3% agar) is also used.

FERMENTATION OF SUGARS

When sugars are fermented, the products of the process are usually alcoholic rather than acid. Great attention is given to the production of carbon dioxide. Sugars commonly used are glucose, galactose, maltose, sucrose, lactose, and raffinose.

The fermentation of the trisaccharide raffinose is used in the identification of many yeasts. The molecule of raffinose is composed of melibiose ($\frac{2}{3}$ of the molecule) and fructose ($\frac{1}{3}$ of the molecule). Van der Walt (1971A) listed data in Table 3.4 as a way to determine the degree of fermentation of the raffinose molecule. *Schizosaccharomyces pombe, Saccharomyces cerevisiae* and *Hansenula anomala* each ferment only $\frac{1}{3}$ of the raffinose molecule.

Sugars are dissolved in a 0.5% solution of commercial powdered yeast extract. A solution of 2% is used for all sugars except raffinose where a 4% concentration is needed. The sugar solutions are put in test tubes containing smaller tubes inverted to collect the carbon dioxide produced by the growing yeasts. In addition to the data in Table 3.4, the following may help to illustrate how raffinose is fermented by yeasts:

(1) CO_2 CO_2 CO_2 No residual sugar
      galactose-glucose-fructose

(2) CO_2 CO_2 ($\frac{2}{3}$) Residual galactose
      galactose-glucose-fructose

(3) CO_2 ($\frac{1}{3}$) Residual melibiose
      galactose-glucose-fructose

Some yeast can ferment galactose alone but not in melibiose (Table 3.4).

TABLE 3.4.   PRESUMED DEGREE FERMENTATION OF THE RAFFINOSE MOLECULE

Sugars				
Raffinose[1]	Melibiose[2]	Sucrose[3]	Galactose	Degree of Fermentation
+	+	+	+	Complete
+	+$\frac{1}{2}$ slowly	+	−	$\frac{2}{3}$ (residual galactose)
+	−	+	+	$\frac{1}{3}$ (residual melibiose)
+	−	+	−	$\frac{1}{3}$ (residual melibiose)
+	+	−	+	$\frac{1}{3}$ (residual sucrose)

Source: Van der Walt (1971A).
[1] A trisaccharide composed of 1 mol of D-galactose, D-glucose, and D-fructose.
[2] A disaccharide composed of 1 mol of D-galactose and D-glucose.
[3] A disaccharide composed of 1 mol of D-glucose and D-fructose.

(4)

$$\text{\underline{galactose-glucose}} \overset{CO_2}{\underset{\uparrow}{\text{-fructose}}}$$

(⅓) Residual melibiose

Some yeast cannot ferment galactose or melibiose.

(5)

$$\overset{CO_2}{\underset{\uparrow}{\text{galactose-}}}\boxed{\text{glucose-fructose}}$$

(⅓) Residual sucrose

OXIDATIVE UTILIZATION OF CARBON COMPOUNDS OR ASSIMILATION TESTS

The auxanographic method of testing, using the yeast-nitrogen base mentioned under media for cultivation (above), is applied here. According to Van der Walt (1971A), the carbon assimilation is more sensitive than fermentation tests for detecting the presence of enzyme systems. In addition to testing sugars (hexose sugars such as glucose, fructose, galactose, raffinose and melibiose), alcohols, acids and their salts, pentose sugars and their derivatives can be used.

IDENTIFICATION OF YEASTS

Morphological characteristics used to identify yeasts are as follows: type of vegetative reproduction, shape and size of cells, ascospore formation, ballistospore formation, shape of the ascospores and ballistopores and macromorphological characteristics of the culture.

Vegetative Reproduction

Vegetative growth is asexual reproduction. *Kloeckera* and *Nadsonia* reproduce by bipolar budding. Multilateral budding is characteristic of *Hansenula, Nematospora, Kluyveromyces, Saccharomyces, Endomycopsis,* and *Candida.*

Yeasts also reproduce by true and pseudomycelium. Example of a yeast reproducing by true mycelium is *Endomycopsis.* One which produces pseudomycelium is *Candida.*

Shape and Size of Cells

The individual yeast cells may be lemon-shaped or apiculate; others are cylindrical, rectangular, oval or round.

Ascospore Formation

The following factors influence ascospore formation:

1. **Condition of the Culture.** Yeasts should be studied as soon as possible after isolation because a culture may lose its ability to form spores.

2. **Temperature.** *Saccharomyces cerevisiae* forms ascospores at a maximum temperature of 35°C and a minimum of 11–12°C. The following cultures sporulated well at room temperature (25°C): *Schizosaccharomyces octosporus* NRRL 854; *Schizosaccharomyces pombe* NRRL 164; *Kluyveromyces fragilis* NRRL 665; *Nadsonia fulvescens* NRRL 991 and *Hansenula saturnus* NRRL 1304. (All strains are from Northern Regional Research Laboratories of USDA.)

3. **Oxygen.** Yeasts need oxygen to sporulate. More ascospores can be found on the outside of yeast cakes than on the inside.

4. **pH.** pH requirements vary. The maximum pH for *Saccharomyces cerevisiae* is 9.1–9.2 and the minimum is 2.4–2.6. For *Schizosaccharomyces pombe* the maximum was 8.0–8.2 with a minimum of 4.0–4.3.

5. **Moisture.** Moisture is necessary for sporulation.

6. **Carbohydrate.** Varies with species.

7. **Nitrogenous Compounds.** The type of nitrogenous compound varies with species.

Ascospore Formation in *Saccharomyces cerevisiae.*—With this yeast, since both a haplophase and a diplophase are perpetuated by budding, they are of equal importance. Normal vegetative cells (alpha or "A" mating types) copulate with each other. When the cells fuse, but before the nuclei of each cell fuse, the condition is called plasmogamy and the gametes have N number of chromosomes (a haploid). The process of nuclear fusion is called karyogamy. The zygote is 2N or a diploid. The diploid may reproduce by budding. The next step, however, in the development of ascospores is to have reduction division with haploid ascospores being formed. If normal somatic cells of alpha and A copulate, they give rise to legitimate diploids which have viable ascospores. However, if two alphas or two A's copulate, two nonviable illegitimate diploids are the result.

Ascospore Formation in *Schizosaccharomyces octosporus.*—In this life cycle, the diploid stage is very short. The zygote undergoes meiosis immediately after karyogamy and develops ascospores. Vegetative cells reproduce by fission. At the proper time, they develop copulation tubes; when two cells come close to each other the cells fuse at these tubes (plasmogamy). The nuclei migrate and fuse (karyogamy). The zygote is 2N. The 2N nucleus undergoes reduction division with the formation of 8 ascospores. The ascospores germinate and repeat the cycle.

Shape of the Ascospores.—Ascospores are found in many forms, among which are hat-shaped, *Hansenula* sp.; Saturn-shaped, *Hansenula saturnus*; needle-shaped, *Nematospora coryli.*

Ballistospore Formation.—Ballistospores are asexual spores formed by budding. Also called blastospores, they occur in the genus *Candida.*

Macromorphological Characteristics of the Culture

As it grows on solid media, the colony may appear rough, smooth, mat or glistening. Some forms, such as *Rhodotorula,* produce pigments. Growth on a liquid may be in the form of pellicle which is characteristic. *Hansenula* produces a pellicle or film on the surface of liquid media. Form and texture of a colony can be reviewed after 4 to 6 weeks if colony is in a screw top bottle to reduce moisture loss.

THE SPOROGENOUS YEASTS OR TRUE YEASTS

The most authoritative work on yeast today is J. Lodder's *The Yeasts, A Taxonomic Study.* This book is edited by Lodder and written by 14 contributors. The book contains 1385 pages. We will follow the classification system as suggested by Lodder (1971A).

Those yeasts which form ascospores are classified in the Class Ascomycetes. These yeasts are also the "true yeasts." After class is the order. The order for the ascosporogenous yeast is Endomycetales (Lodder 1971A). Most of our discussion will be with the Family Saccharomycetaceae. Under the Family Saccharomycetaceae there are 4 subfamilies but we will discuss genera from only 3 of these:

> **Subfamily 1:** Schizosaccharomycoideae with its genus, *Schizosaccharomyces*
> **Subfamily 2:** Nadsonioideae with the one genus *Nadsonia* (There are four genera in this subfamily.)
> **Subfamily 3:** Saccharomycoideae with the genera *Saccharomyces, Pichia, Hansenula, Debaryomyces.*

To show how a classification system works, the following example is given:

Kingdom: Plantae
  Division: Mycota
    Class: Ascomycetes
      Family: Saccharomycetaceae
        Subfamily: Schizosaccharomycoideae
          Genus: *Schizosaccharomyces*
            Species: *octosporus.*

Lodder (1971B) gives the following descriptions for genera under the Class Ascomycetes.

Schizosaccharomyces

The following is quoted from Slooff (1971):

Cells spheroidal to cylindrical, reproducing by fission; true hyphae may develop breaking up into arthrospores.

Asci are produced by somatic conjugation of vegetative cells mostly. Ascospores may be liberated at an early stage. Spores in the asci are globular or oval, seldom oblong. They develop by swelling; during maturation the shape may change to reniform. Infrequently, a rough outer membrane has been observed under the light microscope. A comparison of Fig. 3.1 and 3.2 will show interesting differences in appearance of *Schizosaccharomyces* sp.

Fermentative ability present.

Nitrate not assimilated.

Type species: *Schizosaccharomyces pombe.*

These yeasts have been isolated from cane sugar, cane sugar molasses and palm wine. They have been used to produce African beer as well as industrial alcohol and Jamaican rum. *Schizosaccharomyces* spp. act on lower oligosaccharides but not polysaccharides. They spoil honey, currants, dried prunes, and figs.

S. octosporus has been isolated from honey and raw cane sugar. *S. versatilis* was found in fermenting fruit juice and Portuguese wine.

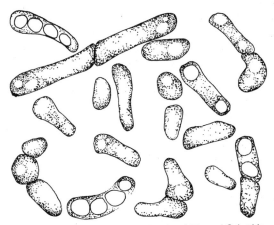

Courtesy of Tom Nelson, University of Missouri-Columbia

FIG. 3.1. *SCHIZOSACCHAROMYCES POMBE* NRRL Y164
After 3 days on yeast extract-malt extract agar.

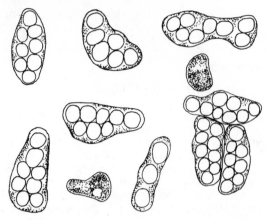

Courtesy of Tom Nelson, University of Missouri-Columbia

FIG. 3.2.   *SCHIZOSACCHAROMYCES OCTOSPORUS* NRRL Y854

After 3 days on yeast extract-malt extract agar.

Nadsonia

The following is quoted from Phaff (1971B):

Cells lemon-shaped, oval, or elongate. Vegetative reproduction by bud-fission at both poles; this process involves the formation of a bud-like structure on a very wide neck. The bud is separated by the formation of a cross wall, followed by fission. Pseudomycelium is not formed; a few chains of elongate cells may be present.

After a heterogamic conjugation between the mother cell and a bud, the contents of the zygote move into another bud formed at the opposite end of the mother cell. This second bud becomes the ascus which is then delimited by a septum; one or more, rarely two, spherical, brownish spiny to warty-walled spores are formed. The spores contain a prominent lipid globule.

In malt extract a pellicle is formed.

One or more sugars are fermented.

Nitrate is not utilized.

No growth at 30°C.

Type species: *Nadsonia fulvescens.*

The yeast was included here because of its interesting cell shape and its ascospore formation as may be seen in Fig. 3.3.

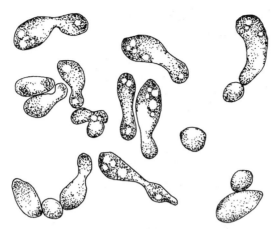

Courtesy of Tom Nelson, University of Missouri-Columbia

FIG. 3.3.   *NADSONIA FULVESCENS* NRRL Y991
After 3 days on yeast extract-malt extract agar.

Saccharomyces

The following is quoted from Van der Walt (1971B):

Vegetative reproduction solely by multilateral budding. Cells spheroidal, ellipsoidal, cylindrical, or elongate. Pseudomycelium may be formed. True mycelium absent.

In liquid media, surface growth may occur after prolonged cultivation. If a pellicle is formed, it is neither pulverulent, dry, nor creeping.

Asci do not rupture on reaching maturity.

Ascospores are spheroidal to prolate-ellipsoidal. Usually one to four ascospores are formed per ascus, on rare occasions more.

Interfertility within certain affinities.

Fermentation.

Lactose and higher paraffins not utilized.

Nitrate not utilized.

Type species: *Saccharomyces cerevisiae.*

Both bread and brewers' yeast are *S. cerevisiae.* They are used in the manufacture of beer, wine, invertase, glycerol and industrial alcohol.

Bottom yeasts remain almost wholly below the surface of liquid during fermentation, forming a sediment in the bottom of the container.

Top yeasts form a frothy scum on the surface of actively-fermenting liquids. After fermentation ceases, the material settles to the bottom of the container. Both top and bottom yeasts are trade terms and are not used in classification.

S. pastorianus produces unpleasant flavors in beer.

S. fragilis is associated with Kefir or fermented milk. These yeasts can produce alcohol from lactose.

S. rouxii and *S. mellis* are osmophilic yeasts and can grow in media containing 60% (W/W) glucose in yeast agar.

The reader is directed to Fig. 3.4, 3.5 and 3.6 to compare *S. cerevisiae, S. rouxii* and *S. uvarum.*

Pichia

The following is quoted from Kreger-van Rij (1971B):

Cells of various shape which reproduce by multilateral budding and in most species form pseudomycelium. True mycelium may occur to a very limited extent.

Iso- or heterogamous conjugation or no conjugation immediately preceding ascus formation.

The spores are spherical, hat- or Saturn-shaped, usually with an oil drop inside. The spores are generally smooth; they may have warts formed exclusively by the outer layer of the spore wall. One to four spores (occasionally more) are formed in each ascus. Usually they are easily liberated from the ascus.

Fermentation absent or present.

Nitrate is not assimilated.

Type species: *Pichia membranaefaciens.*

P. membranaefaciens is a contaminant of beers and wine on which it

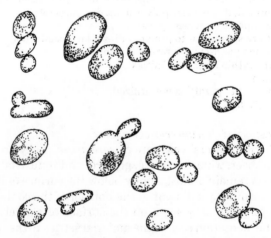

Courtesy of Tom Nelson, University of Missouri-Columbia

FIG. 3.4.   *SACCHAROMYCES CEREVISIAE*
After 3 days on yeast extract-malt extract agar.

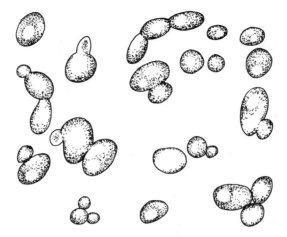

Courtesy of Tom Nelson, University of Missouri-Columbia

FIG.   3.5.   *SACCHAROMYCES   ROUXII* NRRL   Y6681
After 3 days on yeast extract-malt extract agar.

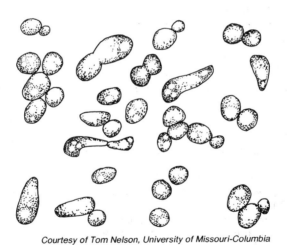

Courtesy of Tom Nelson, University of Missouri-Columbia

FIG.   3.6.   *SACCHAROMYCES   UVARUM* ATCC   9080
After 3 days on yeast extract-malt extract agar.

forms a pellicle. It can occur as a contaminant in the production of food yeasts.

Ethanol can be used as a source of carbon. This yeast can also grow on cucumber brines but grows poorly at salt concentrations above 10%.

Hansenula

The following is quoted from Wickerham (1971):

Asexual reproduction is by budding on a narrow base at various places on the surface of the cell. Cells are spherical, spheroidal, ellipsoidal, oblong, cylindrical or elongated; occasionally, cells that bear long tapers at one or both ends or cells that are markedly elongated to thread-like occur. Pseudohyphae and true hyphae may occur.

Asci have the shape of the vegetative cells and are not rod-shaped as in the nitrate-positive genus *Pachysolen*. From one to four ascospores are produced. Ascospores are hat-shaped, hemispheroidal, spherical or Saturn-shaped. The rings on the latter may be easily seen or may be extremely thin. Ascospores, when observed with the light microscope, have a smooth not a rough surface, as occurs in the nitrate-positive genus *Citeromyces*. Ascospores are usually liberated when mature by rupture of the ascus.

Species of *Hansenula* may be haploid or diploid or have both forms; they may be homothallic or heterothallic. Cultures are long-lived in contrast with cultures of the nitrate-positive genus *Dekkera*.

Sugars may or may not be fermented; pellicles may or may not be formed; esters may be produced; starch is not synthesized.

Nitrate is assimilated.

Type species: *Hansenula anomala.*

Hansenula anomala is used as a food yeast for man. It also causes spoilage of bin-stored rice. Since *H. anomala* is a film yeast, it grows on the surface of cucumber brines when the percentage salt is 10% or less. *H. anomala* occurs on the surface of fruits.

Two species of *Hansenula* are pictured in Fig. 3.7 and 3.8.

Debaryomyces

The following is quoted from Kreger-van Rij (1971A):

Vegetative reproduction by multilateral budding. Various shapes of cell. A primitive or occasionally a well-developed pseudomycelium may be formed. (See Fig. 3.9.)

Heterogamous conjugation, between mother cell and bud, generally precedes ascus formation. Isogamous conjugation also occurs. The spores are spherical or oval. They have warty walls. One to two spores are usually formed per ascus; in some species up to four spores are present in the ascus.

Fermentation rather slow, weak or absent.

Nitrate not assimilated.

Type species: *Debaryomyces hansenii.*

D. membranaefaciens var. *hollandicus* can tolerate 24% sodium chloride. This species forms film on cucumber brines. *D. guilliermondii* var. *novazeelandicus* forms slime on sausages.

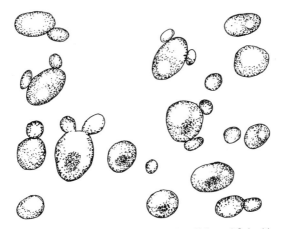

FIG. 3.7.   *HANSENULA ANOMALA* NRRL Y336-8
After 3 days on yeast extract-malt extract agar.

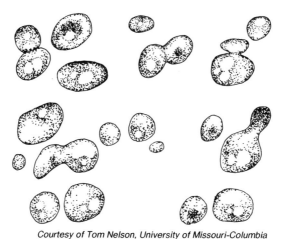

FIG. 3.8.   *HANSENULA SATURNUS* NRRL Y1304
After 3 days on yeast extract-malt extract agar.

Nematospora

The following is quoted from Carmo-Sousa (1971):

Cells of various shapes reproducing vegetatively by multipolar budding.
Well-developed pseudomycelium usually present. True mycelium composed
of sparsely septate, branched hyphae. No arthrospores.

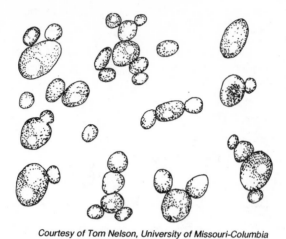

Courtesy of Tom Nelson, University of Missouri-Columbia

FIG. 3.9.   *DEBARYOMYCES KLOECKERI* NRRL Y1449
After 3 days on yeast extract-malt extract agar.

Sporulating cells elongate, larger than the vegetative cells, dehiscing early after attaining maturity.

Spores spindle-shaped, provided with a whip-like appendage; usually eight per cell arranged in two bundles of four. Occasionally less or more than eight spores per cell. Many spores apparently septate near the equator of the spindle.

Alcoholic fermentation present.

Nitrate not assimilated.

Type species: *Nematospora coryli.*
N. coryli causes rots in citrus fruits.

THE ASPOROGENOUS YEASTS OR THE FALSE YEASTS

These are classified in the Class Deuteromyces or Fungi Imperfecti.

According to Lodder (1971A), these false yeasts (those which do not form ascospores) are classified in the order Moniliales and the family Cryptococcaceae.

Candida

The following is quoted from Van Uden and Buckley (1971):

Cells globose; ovoid, cylindrical, or elongate, sometimes irregularly shaped, normally not ogival, apiculate or flask-shaped.

Reproduction normally by multipolar budding; cells with apparent bipolar budding do not normally bud on a broad base.

Pseudomycelium formation by all or most strains of all species and varieties. Pseudomycelium is often differentiated into pseudohyphae and blastospores. Chlamydospores may be formed. True mycelium may be formed. Arthrospores absent. Ascospores, teliospores or ballistospores not formed.

Visible pigmentation due to carotenoid pigments absent.

Extracellular polysaccharides may be formed and may give a positive iodine reaction.

Alcoholic fermentation occurs in many species.

The reader is directed to Fig. 3.10 and 3.11 to note differences in morphology.

Type species: *Candida tropicalis.*

C. tropicalis and *C. utilis* are used as a food yeast for man, poultry, pigs and dogs. *C. tropicalis* spoils rice oil. *C. vini* may occur in breweries and can frequently be isolated from beer. They grow as a film on liquids where there is ample oxygen available.

Kloeckera

The following is quoted from Phaff (1971A):

Cells lemon-shaped (apiculate), ovoidal, sausage-shaped or irregular elongate. Reproduction by bipolar budding on a moderately broad base. A pseudomycelium is usually not formed, but some strains produce a pseudomycelium ranging from primitive to well developed. Ascospores are not formed.

Fermentation of sugars occurs; nitrate is not assimilated; all species have an absolute requirement for inositol and pantothenic acid.

Type species: *Kloeckera apiculata.*

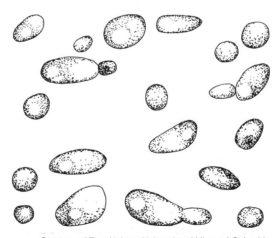

Courtesy of Tom Nelson, University of Missouri-Columbia

FIG. 3.10.   *CANDIDA UTILIS* ATCC 9950
After 3 days on yeast extract-malt extract agar.

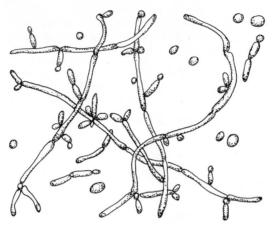

Courtesy of Tom Nelson, University of Missouri-Columbia

FIG. 3.11.   *CANDIDA TROPICALIS* ATCC 9968
After 3 days on yeast extract-malt extract agar.

K. apiculata occurs in the fermentation of grape juice but is eliminated as *Saccharomyces cerevisiae* produces alcohol. This species is common on ripe strawberries, cherries, plums, grapes, and gooseberries.

Rhodotorula

The following is quoted from Phaff and Ahearn (1971):

> Cells spheroidal, ovoidal or elongate. Reproduction by multilateral budding. Occasional strains of some species may form chlamydospore-like cells and (or) pseudo- or true hyphae of various lengths. Ascospores or ballistospores are not formed. Red and (or) yellow carotenoid pigments are synthesized in young malt agar cultures.
>
> No assimilation of inositol as sole source of carbon. Starch-like substances are not synthesized. Fermentative ability is lacking.
>
> Many strains have a mucous appearance due to capsule formation, but others are pasty or dry and wrinkled.
>
> Acid formation on chalk agar and gelatin liquefaction are generally negative, except for a weakly positive reaction in certain strains of some species.

Examples of multilateral budding in *Rhodotorula* sp. may be seen in Fig. 3.12.

Type species: *Rhodotorula glutinis.*

R. glutinis has been used to produce fat. *Rhodotorula* sp. contaminates beef, dairy products and kraut.

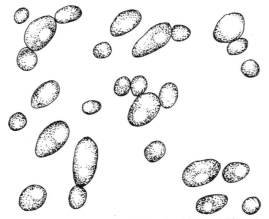

Courtesy of Tom Nelson, University of Missouri-Columbia

FIG. 3.12. *RHODOTORULA* SP.
After 3 days on yeast extract-malt extract agar.

KEY TO THE GENERA TO BE STUDIED IN THE LABORATORY

In *The Yeasts, A Taxonomic Study* by Lodder (1971B), a key used for identification of 39 genera of yeast is included; the portion which follows is for 8 genera suggested for study in the laboratory.

Key to Genera

A. Vegetative reproduction by fission without constriction of the cell, *Schizosaccharomyces*   41

B. Vegetative reproduction by bipolar budding
    Ascospores formed, *Nadsonia*   42
    Ascospores not formed, *Kloeckera*   49

C. Vegetative reproduction by multilateral budding; true mycelium, arthrospores, ballistospores may also occur. Ascospores formed.
    1.   No true mycelium or scarce and budding cells
       Ascospores hat- or Saturn-shaped, *Hansenula*   46
       Ascospores whip-like, *Nematospora*   47
       Ascospores spherical or oval, *Saccharomyces*   43
    2.   Abundant true mycelium and budding cells,
       Endomycopsis   [1]

D. Vegetative cells and pseudomycelium always present; true mycelium may be formed; no ascospores formed, *Candida*   48

[1] Description not given herein.

PHYSIOLOGICAL CHARACTERISTICS USED TO CLASSIFY YEASTS

For the most part, yeasts can be keyed to genera by morphological characteristics as illustrated above. In the further separation into species, more emphasis is put on physiological properties. The following are characteristics used by Van der Walt (1971A):

1. Utilization of carbon compounds.
   a. Fermentative utilization of carbon compounds.
   b. Oxidative or assimilative utilization of carbon compounds.
   c. Splitting of arbutin.
2. Utilization of nitrogen compounds.
3. Growth in vitamin-free medium.
4. Growth on media of high osmotic pressure.
5. Growth at elevated temperatures.
6. Acid production.
7. Production of extracellular, amyloid compounds.
8. Hydrolysis of urea.
9. Fat-splitting.
10. Pigment formation.
11. Ester production.
12. Resistance to actidione.
13. Gelatin liquefaction.

CHEMICAL COMPOSITION OF YEAST CELLS

The chemical composition of bakers' and brewers' yeasts is listed in Table 3.5. Yeast is a good source of protein, but the RNA content of the cell limits the amount to no more than 2 g of RNA per day. There are heat-shock methods to activate the RNAase and lower the RNA content to about 1%. The ingestion of excessive amounts of microbial RNA causes the human body to develop gout.

CYTOLOGY OF YEAST

The yeast cell consists of a cell wall, plasma membrane, cytoplasm, centrosome, centrochromatin, nuclear membrane, mitochondria, nucleolus, and vacuoles (Salle 1973). See Table 3.6 for explanation of function and nature of structures.

TABLE 3.5.   CHEMICAL COMPOSITION OF BREWERS' YEAST

Component	Amount
Protein	176 g/454 g
Fat	4.5 g/454 g
Carbohydrate	174.2 g/454 g
Thiamin	70.81 mg/454 g
Riboflavin	19.41 mg/454 g
Niacin	171.9 mg/454 g
Isoleucine	2267 mg/100 g
Leucine	3105 mg/100 g
Lysine	3509 mg/100 g
Methionine	621 mg/100 g
Cystine	348 mg/100 g
Total S-containing AA	969 mg/100 g
Phenylalanine	1882 mg/100 g
Tyrosine	1608 mg/100 g
Total aromatic AA	3490 mg/100 g
Threonine	2149 mg/100 g
Valine	2850 mg/100 g
Arginine	1944 mg/100 g
Histidine	969 mg/100 g
Alanine	2621 mg/100 g
Aspartic acid	4210 mg/100 g
Glutamic acid	4154 mg/100 g
Glycine	1863 mg/100 g
Proline	1497 mg/100 g

Source: FAO (1970) and Watt and Merrill (1963).

TABLE 3.6.   STRUCTURE AND SOME FUNCTIONS OF YEAST CELL PARTS

Structure	Function and Nature
Cell wall	About 1.2 nm thick and occupies 40% of its volume. Composed of yeast cellulose and chitin. Gives cell its rigid structure.
Cytoplasmic or plasma membrane	Determines substances entering and leaving cell.
Mitochondria	Contain a small amount of DNA with larger amounts of RNA. Spherical rod-shaped or thread-like structures. Site of cellular respiration, energy-producing process.
Centrosome	Plays a leading role in budding, copulation, and meiosis.
Centrochrom-atin	Structure is always attached to external surface of centrosome.
Nucleus	Mainly contains DNA with RNA in smaller amounts.
Vacuoles	Storage place of various chemicals, especially glycogen.
Metachromatic granules	Storage place of phosphate.
Fat globules	Reserve material in the cytoplasm.

Source: Salle (1973).

RELATION OF SOME YEASTS IN THE CLASS ASCOMYCETES TO YEASTS IN THE CLASS DEUTEROMYCETES

The relationship between *Endomycopsis, Candida, Saccharomyces, Hansenula,* and *Cryptococcus* has been made by Skinner *et al.* (1947).

Endomycopsis forms both true mycelium and budding single cells. Ascospores are formed by the fusion of contiguous cells. If this yeast lost its ability to form ascospores, it would give rise to some *Candida* species which form true mycelium and single budding cells.

However, if *Endomycopsis* lost its ability to form true mycelium, it would become like *Saccharomyces* and *Hansenula*. Both of these genera produce ascospores but no mycelium. If *Hansenula* and *Saccharomyces* lost their ability to form spores, they would become like yeasts in the genus *Cryptococcus*. One may use such comparisons to show the evolution of various genera of yeasts.

PRESERVATION OF YEAST CULTURES

Some yeasts may be preserved for a short time (1 year) on agar slants. Malt or wort agar may be used. Slants should be prepared in screw-top test tubes to ensure that the cultures do not dry out. Cultures should be refrigerated at 4°C.

Yeast cells may be lyophilized (freeze-dried) in skimmed milk (10%). The tubes should be sealed under vacuum and stored at 4°C until opened and grown out in malt extract or wort broth.

Yeast cultured on agar slants may be preserved under mineral oil for as long as 6 years at 4°C. There should be abundant growth on the slant and the oil should be at a level of at least 1 cm above the tip of the agar.

GLOSSARY

Apiculate—Lemon-shaped.

Arthrospore—A spore resulting from the fragmentation of hypha. Thin-walled cells.

Asci—Plural of ascus.

Ascospores—Sexual spores produced in an ascus.

Ascus—A sac-like structure containing ascospores. Ascospores result from karyogamy and meiosis.

Asporogenous—Without spores or no spores are produced.

Assimilation—To take up and incorporate into cell cytoplasm. Aerobic.

Ballistospore—Spores produced on sterigmata that protrude from vegetative cells and are ejaculated into the air by the so-called drop mechanisms.

Bipolar budding—Budding only at the poles or ends of the cells.

Blastospore—A bud or spore.

Budding—A small outgrowth from the mother cell which expands into a new individual. A form of asexual reproduction.

Chlamydospore—A thick-walled spore in a hyphal cell acting as a resting spore.

Conjugation and copulate—The fusion of cells followed by the fusion of nuclei to form a zygote.

Diploid—2N number of chromosomes.

Diplophase—Cells containing 2N number of chromosomes.

Fermentation—Changing of sugars into alcohol and carbon dioxide. Anaerobic.

Fission—Reproduction by splitting in two to form two cells.

Haploid or haplophase—That part of the life cycle which is haploid (N). N number of chromosomes is the reduced number. See diploid and diplophase.

Heterogamous—Having male and female cells which are distinguishable morphologically.

Isogamous—Gametes which are indistinguishable morphologically.

Karyogamy—The fusion of two nuclei.

Meiosis—A pair of nuclear divisions in quick succession with one being reductional (from 2N to N).

Multilateral budding—Budding from all sides of the cell.

Multipolar—Budding from all sides of the cell.

Osmophilic—Loving an environment of high osmotic pressure.

Pellicle—A film on a liquid.

Plasmogamy—The fusion of two protoplasts.

Pseudohyphae—Same as pseudomycelium, except less extensive in structure.

Pseudomycelium—A false mycelium or where cells adhere end to end after dividing. Occurs in some yeasts.

Reniform—Bean-shaped.

Sporogenous—Produces spores.

True mycelium—A mass of hyphae constituting the body of a fungus where the individual cells are not formed by budding.

Zygote—A diploid cell resulting from the union of two haploid cells. The nuclei fuse (2N).

REVIEW QUESTIONS

1. Discuss the occurrence and the importance of yeasts to man.
2. How are yeasts isolated from nature?
3. What media are used for cultivation of yeasts?

4. How does one test for the fermentation of a sugar? The assimilation of a carbon compound?
5. Describe and draw the types of asexual reproduction in the yeast.
6. Draw different kinds of ascospores. Identify the kind of ascospore with the genus in which it occurs.
7. What is a ballistospore? Do they occur in genera we are studying?
8. List the genera in the Class Ascomycetes which we are studying.
9. In how many classes are yeasts classified?
10. Make a list of yeasts that cause food spoilage.
11. What yeasts are important industrially?
12. What are the main characteristics that are used to separate genera in the identification of yeasts?
13. List physiological characteristics used to classify yeast.
14. Draw a yeast cell and label its internal parts.
15. Discuss how to preserve yeast cultures.
16. What yeasts (which we are studying) form films?
17. What is meant by top and bottom yeasts?
18. Be able to define all terms in the glossary.
19. Show by words and drawings the life cycle of *Saccharomyces cerevisiae* and *Schizosaccharomyces octosporus*.
20. Explain the relationship between *Endomycopsis, Candida, Saccharomyces* and *Cryptococcus*.

LABORATORY EXERCISES

The representative yeasts listed in Table 3.7 were selected for laboratory study. A suggested approach follows.

Streak the mixture of yeasts on malt agar. Incubate plates at room temperature. Identify the isolated yeasts to genera by the key in this chapter.

Draw the shapes of each isolate. Stain smears with malachite green and counter-stain with safranin. Good contrasts between red vegetative cells and green ascospores are observed in stained preparations of *S. uvarum, K. fragilis,* and *S. cerevisiae*. Careful examination of slides of *S. pombe* and *S. octosporus* is needed to differentiate the red and the green staining. In young cultures, the ascospores of *N. fulvescens* show good contrast. In old cultures, the ascospores become free of the vegetative cell.

If young cultures of *N. coryli* are stained, they will have eight ascospores per ascus. The cell wall of the ascus does not stain well. In old cultures, individual ascospores appear free on the slides.

TABLE 3.7. LIST OF YEASTS TO BE STUDIED IN THE LABORATORY

Yeast	Economic Importance
Hansenula anomala NRRL Y 1304	Causes ester-like taste in beer.
Saccharomyces uvarum ATCC 9080	Used to produce beer; used as food; produces ergosterol; this yeast can be used to assay for pantothenic acid and pyridoxine.
Nadsonia fulvescens NRRL Y 991	Isolated from sap of trees; no known economic importance.
Hansenula saturnus NRRL Y 1304	Film yeast on kraut and pickles.
Kluyveromyces fragilis NRRL Y 665	Ferments lactose; ferments milk products; converts dairy wastes into yeast cells.
Schizosaccharomyces pombe NRRL Y 164	Used to make rum.
Schizosaccharomyces octosporus NRRL Y 854	Grows in sugar cane juice.
Rhodotorula sp.	Grows on meat, dairy products, kraut.
Nematospora coryli NRRL 1343	Causes diseases of plants.
Saccharomyces rouxii NRRL Y 6681	Osmophilic yeast spoiling jellies, jams, and dried fruit.
Saccharomyces cerevisiae	Used in baking and alcoholic beverages.
Endomycopsis fibuligera ATCC 9947	Produces amylases (alpha and beta).

REFERENCES

BBL. 1968. BBL Manual of Products and Laboratory Procedures, 5th Edition. BBL, Division of Becton, Dickinson and Co., Cockeysville, Maryland.

DIFCO LABORATORIES. 1953. Difco Manual of Dehydrated Culture Media and Reagents for Microbiological and Chemical Laboratory Procedures, 9th Edition. Difco Laboratories, Detroit.

CARMO-SOUSA, L. 1971. Genus 13, *Nematospora. In* The Yeasts, A Taxonomic Study. J. Lodder (Editor). North-Holland Publishing Co., Amsterdam, The Netherlands.

FAO. 1970. Amino Acid Content of Foods. Food and Agriculture Organization of the United Nations, Rome, Italy.

INGRAM, M. 1958. Yeast in food spoilage. *In* The Chemistry and Biology of Yeasts. A. H. Cook (Editor). Academic Press, New York.

KREGER-VAN RIJ, N. J. W. 1971A. Genus 3. *Debaryomyces. In* The Yeasts, A Taxonomic Study. J. Lodder (Editor). North-Holland Publishing Co., Amsterdam, The Netherlands.

KREGER-VAN RIJ, N. J. W. 1971B. Genus 15. *Pichia. In* The Yeasts, A Taxonomic Study. J. Lodder (Editor). North-Holland Publishing Co., Amsterdam, The Netherlands.

LODDER, J. 1971A. General classification of yeasts. *In* The Yeasts, A Taxonomic Study. J. Lodder (Editor). North-Holland Publishing Co., Amsterdam, The Netherlands.

LODDER, J. 1971B. Introduction to the chapters IV, V, VI and VII and Key to genera. *In* The Yeasts, A Taxonomic Study. J. Lodder (Editor). North-Holland Publishing Co., Amsterdam, The Netherlands.

LUND, A. 1958. Ecology of yeasts. *In* The Chemistry and Biology of Yeasts. A. H. Cook (Editor). Academic Press, New York.

PHAFF, H. J. 1971A. Genus 4. *Kloeckera*. *In* The Yeasts, A Taxonomic Study. J. Lodder (Editor). North-Holland Publishing Co., Amsterdam, The Netherlands.

PHAFF, H. J. 1971B. Genus 12. *Nadsonia*. *In* The Yeasts, A Taxonomic Study. J. Lodder (Editor). North-Holland Publishing Co., Amsterdam, The Netherlands.

PHAFF, H. J. and AHEARN, D. G. 1971. Genus 7. *Rhodotorula*. *In* The Yeasts, A Taxonomic Study. J. Lodder (Editor). North-Holland Publishing Co., Amsterdam, The Netherlands.

SALLE, A. J. 1973. Fundamental Principles of Bacteriology. McGraw-Hill Book Co., New York.

SKINNER, C., EMMONS, C. W., and TSUCHIYA, H. M. 1947. Henrici's Molds, Yeasts, and Actinomycetes. John Wiley & Sons, New York.

SLOOFF, W. C. 1971. Genus 19. *Schizosaccharomyces*. *In* The Yeasts, A Taxonomic Study. J. Lodder (Editor). North-Holland Publishing Co., Amsterdam, The Netherlands.

VAN DER WALT, J. P. 1971A. Criteria and methods used in classification. *In* The Yeasts, A Taxonomic Study. J. Lodder (Editor). North-Holland Publishing Co., Amsterdam, The Netherlands.

VAN DER WALT, J. P. 1971B. Genus 16. *Saccharomyces*. *In* The Yeasts, A Taxonomic Study. J. Lodder (Editor). North-Holland Publishing Co., Amsterdam, The Netherlands.

VAN UDEN, N. and BUCKLEY, H. 1971. Genus 2. *Candida*. *In* The Yeasts, A Taxonomic Study. J. Lodder (Editor). North-Holland Publishing Co., Amsterdam, The Netherlands.

WATT, B. K. and MERRILL, A. L. 1963. Composition of Foods. USDA Handbook 8, Washington, D.C.

WICKERHAM, L. J. 1971. Genus 7. *Hansenula*. *In* The Yeasts, A Taxonomic Study. J. Lodder (Editor). North-Holland Publishing Co., Amsterdam, The Netherlands.

4

Molds

OCCURRENCE AND IMPORTANCE

Plant Pathogens

Many plant pathogens attack fruits and vegetables in the field and cause decay during transportation and storage. For the most part, however, plant pathogens will not be studied in this class. One exception, however, is *Phytophthora infestans* which attacks potato and tomato plants as well as the potato tuber and tomato fruit. The blight that this fungus causes on potatoes was responsible for the Irish migration to the United States. The blight struck, causing a famine in Ireland during 1845–1846 (Chester 1947).

Another fungus considered to be a plant pathogen is *Monilinia fructicola*. This fungus has a life cycle which shows the relation between the Class Ascomycetes and Fungi Imperfecti. The perfect stage of most Fungi Imperfecti has not been isolated. The fungus causes brown rot of stone fruits. As much as 40% of the peach crop has been lost in some years due to brown rot (Chester 1947).

Food Spoilage During Storage

Since spores of the common molds occur in the air, dust and soil, growth will occur on a suitable substrate which has been exposed and then incubated at room temperature (25°C). The filamentous fungi which we call molds and with which we deal in food microbiology are saprophytes. Since they can produce billions and billions of spores or conidia on organic matter (nonliving), they find their way into all parts of our environment. Because of their extremely small size, the conidia are carried easily by wind currents.

59

If appropriate moisture and temperature are available, these microorganisms attack our foods during storage.

Billions and billions of dollars worth of food and fiber are lost each year due to spoilage around the world. With the ever-increasing world population to feed, such losses must be brought to the lowest level possible. Too much food is lost after harvest but before consumption in the United States (USDA 1965).

Enzyme Production

Amylases for saccharifying starches are produced by *Aspergillus oryzae, Mucor rouxii, Rhizopus delemar* and *Rhizopus japonicus* (Prescott and Dunn 1949). *Aspergillus oryzae*, an industrially important mold, produces amidase, amylase, catalase, cytase, dextrinase, emulsin, α-glucosidase, β-glucosidase, histozyme, inulase, invertase, lactase, lecithinase, lipase, maltase, protease, rennet, sulphatase and tannase (Prescott and Dunn 1949).

Production of Specific Foods and Ingredients

Molds have been used to produce citric acid, gluconic acid, fumaric acid, lactic acid and fat. Molds are also used in production of cheeses. The white mold, *Penicillium camemberti*, secretes proteolytic enzymes into the curd, causing it to become soft and to develop a mild flavor which is characteristic of Camembert cheese. However, as acidity decreases, bacteria and *Geotrichum candidum* produce an ammoniacal flavor during the ripening process.

Another cheese in which a mold is used is Roquefort cheese. During the ripening period, the fungus *Penicillium roqueforti* produces lipolytic and proteolytic enzymes responsible for the pungent flavor of Roquefort cheese. The characteristic taste is due to the breakdown of fat and the resulting production of capric, caproic and caprylic acids (Moore-Landecker 1972).

Rhizopus oligosporus grows on dehulled and boiled soybeans for 24 hr to produce tempeh. During this time, the fungus penetrates the soybean, producing large quantities of proteolytic enzymes. As a result, the soybeans are more digestible and have an increase in riboflavin, niacin, and B_{12} (Moore-Landecker 1972).

Aspergillus oryzae is used to produce shoyu or soya sauce, a product that is consumed in Japan and in other countries around the world.

Molds as Assay Microorganisms

Molds are used for vitamin analysis. *Phycomyces blakesleeanus* can be used for the assay of thiamin and *Neurospora crassa* to test for *p*-aminobenzoic acid and inositol. The amounts of biotin present can be determined with *Neurospora crassa* and *Neurospora sitophila*. *Aspergillus niger* will respond to different amounts of copper, magnesium, potassium and molybdenum (Lilly and Barnett 1951).

Human Pathogens

Some filamentous fungi are pathogens on man but these will not be covered in this book.

ISOLATION PROCEDURES

The isolation procedures (Wolf and Wolf 1947) for fungi are similar to those methods used in isolating bacteria and yeasts.

Surface Disinfection.—In isolating fungi from diseased or decaying plant material, it is a good idea to surface disinfect to eliminate contaminants which are not causing the disease or spoilage. Such disinfection may be accomplished by any of the methods which follow:

1. Using 95% alcohol for a few seconds; removing by flaming or washing with sterile distilled water.
2. A 1:1000 solution of $HgCl_2$ applied 15 to 45 sec and removed by washing with sterile distilled water.
3. A solution of calcium or sodium hypochlorite applied for about 1 min and removed by washing with sterile distilled water.
4. A solution of 50% H_2O_2 applied for 15 sec to 5 min and removed by washing with sterile, distilled water.

The material is cut open, and plant tissue is taken from the edge of the sound and decaying matter. This is transferred aseptically to a suitable growth medium and incubated at room temperature.

If there are enough spores present, isolation can be achieved by the classical streaking or dilution method with poured plates. Undesirable fungal and bacterial contaminants may cause difficulties in isolating a pure culture. The following are ways to overcome these problems:

1. The addition of oxgall inhibits some bacteria and restricts spread of the mold colony.

2. Crystal violet may be added to eliminate many Gram-positive bacteria, and streptomycin may be added to kill the Gram-negative cells.
3. Adjustment of the growth media to a pH of 3.5 will eliminate many bacteria. However, acid agar also reduces the total number and kinds of molds.
4. Rose bengal has also been used as a bacterial inhibitor.

MEDIA FOR CULTIVATION

There are several media that may be used to cultivate molds. The ones which permit growth of yeast (without adding an inhibitor for molds) will allow development of filamentous fungi. Three examples are potato-dextrose, malt, and wort agars. Since fungi vary in appearance with the substrate, Czapek's solution agar is used to grow unknowns of aspergilli and the penicillia. The composition of this medium is (taken from p. 102, BBL, 1968):

	g/liter
$NaNO_3$	3.0
K_2HPO_4	1.0
$MgSO_4 \cdot 7H_2O$	0.5
KCl	0.5
$FeSO_4 \cdot 7H_2O$	0.01
Sucrose	30.0
DW	(1000 ml)
Agar	15.0

GENERAL CONSIDERATIONS

A fungus is a plant which has no roots, stems, leaves or chlorophyll; its vegetative structure is a thallus, an undifferentiated plant body with all cells essentially alike. The vegetative phase consists of developing mycelium and produces by-products which are important in industry or in the spoilage of foods. The reproductive or the conidial stage is the contaminating phase.

A mold's vegetative structure or soma consists of branching tubes with or without walls forming individual cells. This network of filaments (hyphae, singular hypha) is called the mycelium. Two special types of hyphae are rhizoids (a short and thin branched thallus resembling a root) and stolons (hyphae connecting two groups of rhizoids). Growth occurs at the tip of the hypha, but food material may be absorbed over the entire surface of the mycelium. With some pathogenic fungi which attack plants, the fungus obtains its food by special hyphae, haustoria, which are inserted into a host cell.

There are some fungi which have cross walls (septate mycelium) only when producing reproductive structures. Their mycelium is considered non-septate or coenocytic. Coenocytic refers to the fact that nuclei are embedded in the cytoplasm without being separated by cross walls. The absence of septa in the hyphae of the two classes (Phycomyces) Oömycetes and Zygomycetes serves to separate these fungi from the Ascomycetes.

Most of the filamentous fungi we will study are septate and belong to the Fungi Imperfecti (fungi not producing sexual spores). There is one fungus, an ascomycete, which causes brown rot of peaches and which shows us the relation of some Ascomycetes to the Fungi Imperfecti.

The movement of cytoplasm within the hyphal cells can be observed. This movement, as seen under the microscope, is called cytoplasmic streaming. The cytoplasm has a granular (or foamy) appearance due to granules and vacuoles. Reserve materials are fat, carbohydrate, and proteins. The carbohydrate is stored as glycogen.

Growth of mycelium may be extensive. Every 30–40 min a new hypha may be produced. If all hyphae were laid end to end, the fungus might accumulate a distance of $\frac{1}{2}$ mile in 24 hours and hundreds of miles in 48 hours.

Some of the fungi (*Aspergillus* and *Penicillium*, for example) form a pellet or hard mass of hyphae (sclerotia) which is resistant to heat, cold and dryness. These sclerotia may be more or less spherical. Fungi also produce a strand of hyphae which functions like sclerotia. These strands of hyphae are called rhizomorphs. Both help to preserve the fungus.

SPECIAL KINDS OF ASEXUAL REPRODUCTION

Chlamydospores. These are thick-walled spores produced in mycelium or in conidia.

Conidia. Spores produced on a conidiophore not in an enclosed structure.

Sporangiospore. Spores produced in a sac-like structure called a sporangium.

Oidia. Thin-walled spores (also known as arthrospores) produced by fragmentation of hyphae.

SEXUAL REPRODUCTION

The following are used as examples and do not cover all ways of sexual reproduction in fungi. They were selected to illustrate types of sexual reproduction in types of molds encountered by food microbiologists.

Class Oömycetes

Genus *Phytophthora*. In the species *Phytophthora infestans*, there are two structures involved in sexual reproduction. The female form is the oögonium and the male is the antheridium. When these structures fuse, there is migration of male nuclei into the oögonium. All but one nucleus of the antheridium and oögonium disintegrate after fusion (a process called karyogamy). An oöspore, or zoospore, is formed from this union (2N). At the germination of the oöspore, reduction division takes place. Normal asexual reproduction of this fungus is by sporangiospores.

Sexual reproduction in the genus *Pythium* is similar.

Class Zygomycetes

Genus *Rhizopus*. In the fungus *R. nigricans* (*R. stolonifer*), sexual reproduction takes place between positive and negative mating strains. These strains produce special cells called suspensors which fuse to form a zygote (2N). Upon germination of the zygote, meiosis occurs, giving rise to (N) mycelium. (Meiosis is a special type of nuclear division. A pair of nuclear divisions occurs in quick succession, one of which is reductional. This is contrasted with mitosis which is nuclear division without reduction.) All species are heterothallic.

Class Ascomycetes

Genus *Monilinia* (*Sclerotinia*). Asci are produced in an open ascocarp called an apothecium.

Genus *Eurotium*. Asci are produced in a closed ascocarp.

Genus *Neurospora*. Asci are produced in a closed ascocarp with a pore at the top. The ascocarp is flask-shaped.

GENERA CLASSIFIED BY ASEXUAL CHARACTERS

Sexual reproduction occurs more rarely than does the asexual. It is the asexual stage which is active in contamination. Identifications of the fungi are usually based upon the asexual methods of reproduction.

Genera That Are Separated by the Methods of Asexual Reproduction in the Class Zygomycetes

Genus *Absidia*

a. There are stolons ("runners").
b. Sporangiophores occur in the middle of the stolons or at the internodes.
c. The sporangium is pear-shaped.
d. The sporangiophores are septate.
e. There is an apophysis at the base of the sporangium.

Genus *Rhizopus*

a. There are stolons.
b. Sporangiophores always occur in groups at the same node.
c. Rhizoids occur at the same node.
d. Sporangia are spherical. A typical sporangium is shown in Fig. 4.1.
e. Sporangiophores are unbranched.
f. Sporangiospores are striated and oblong.

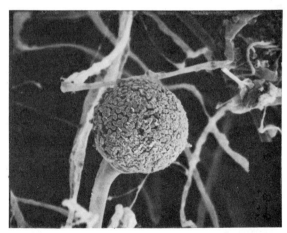

Courtesy of M. F. Brown and H. G. Brotzman, University of Missouri-Columbia

FIG. 4.1. SPORANGIUM OF *RHIZOPUS STOLONIFER* X400

Genus *Mucor*

a.   There are no stolons or rhizoids.
b.   Sporangiophores arise singly unbranched or branched.
c.   Sporangia are spherical.
d.   Sexual reproduction is the same as that of *Rhizopus*.
e.   Some species are heterothallic; others are homothallic.

Genus *Thamnidium*

a.   There is no columella.
b.   There are sporangioles (small sporangia with 6 to 8 spores per sporangium).
c.   Sporangium at the top of the sporangiophore contains numerous spores.

Asexual Reproduction in the Class Deuteromycetes (Fungi Imperfecti)

This class has no sexual means of reproduction except in special cases where they are related to Ascomycetes. The following descriptions of genera in this class were taken from Barnett's (1960) *Illustrated Genera of Imperfect Fungi.*

Genus *Alternaria.* Conidiophores dark, simple, rather short or elongate, typically bearing a simple or branched chain of conidia; conidia dark,

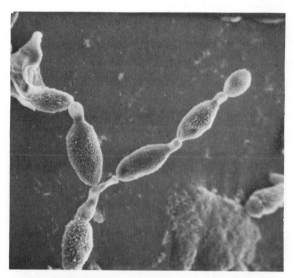

Courtesy of M. F. Brown and H. G. Brotzman, University of
Missouri-Columbia

FIG. 4.2.   CONIDIA OF *ALTERNARIA CITRI* ×1500

typically with both cross and longitudinal septa; variously shaped, obclavate to elliptical or ovoid, frequently borne acropetally in long chains, less often borne singly and having an apical simple or branched appendage; parasitic or saprophytic on plant material.

A. tenuis produces conidia in chains; *A. solani* produces conidia singly. Conidia of *A. citri* occur in branched chains as illustrated in Fig. 4.2.

Genus *Fusarium*. Mycelium extensive and cottony in culture, often with some tinge of pink, purple or yellow, in the mycelium or medium; conidiophores variable, slender and simple, or stout, short, branched irregularly or bearing a whorl of phialides, single or grouped into sporodochia; conidia, hyaline, variable, principally of two kinds, often held in a mass of gelatinous material; macroconidia several-celled slightly curved or bent at the pointed ends, typically canoe-shaped; microconidia 1-celled, ovoid or oblong, borne singly or in chains; some conidia intermediate 2- or 3-celled, oblong or slightly curved; parasitic on higher plants or saprophytic on decaying plant material.

Typical spores, or macroconidia, may be seen in Fig. 4.3.

Genus *Sporotrichum*. Conidiophores hyaline, sometimes simple, usually irregularly branched, with spore bearing portion near the apex; conidia hyaline, 1-celled, globose or ovoid, attached apically and laterally; saprophytic in soil, parasitic on higher plants or pathogenic on animals.

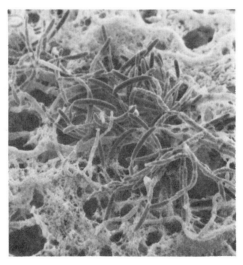

Courtesy of M. F. Brown and H. G. Brotzman, University of Missouri-Columbia

FIG. 4.3. MACROCONIDIA OF *FUSARIUM SO-LANI* ×750

Genus Botrytis. Conidiophores long, slender, often pigmented, branched, the apical cells enlarged or rounded, bearing clusters of conidia on short sterigmata; conidia hyaline or ash-colored, gray in mass; 1-celled, ovoid; black irregular sclerotia frequently produced; parasitic, causing gray mold on many plants, or saprophytic.

The grape-like clusters of conidia, as in Fig. 4.4, give rise to the name *Botrytis.*

Genus Monilia. Mycelium white or gray, abundant in culture; conidia pink, gray, or tan in mass, 1-celled, short cylindric to rounded, catenulate, formed acropetally; conidiophore branched, its cells differing little from the older conidia. Some species are imperfect stages of *Neurospora* and are common saprophytes; others, whose perfect states are *Monilinia* (*Sclerotinia* spp.) cause brown rots of fruits.

Courtesy of M. F. Brown and H. G. Brotzman, University of Missouri-Columbia

FIG. 4.4.   CONIDIA OF *BOTRYTIS CINEREA* ×500

The asexual spores of *M. fructigena* (conidia) which are shown in Fig. 4.5, are responsible for the rapid spread of the pathogen in the summer.

Genus Geotrichum. Mycelium white, septate; conidia (oidia) hyaline, 1-celled, short cylindrical with truncate ends, formed by segmentation of hyphae; saprophytic, common in soil.

Figure 4.6 illustrates the distinctive shape of the conidia.

Genus Aspergillus. Conidiophores upright, simple, terminating in a globose or clavate swelling, bearing phialides at the apex or radiating from

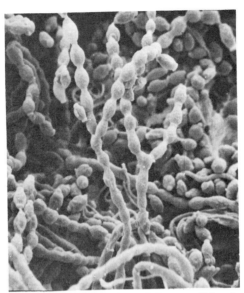

Courtesy of M. F. Brown and H. G. Brotzman, University of Missouri-Columbia

FIG. 4.5.   CONIDIA OF *MONILIA FRUCTIGENA*
×750

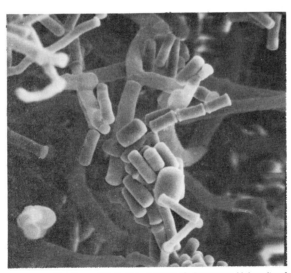

Courtesy of M. F. Brown and H. G. Brotzman, University of Missouri-Columbia

FIG. 4.6.   CONIDIA (OIDIA) OF *GEOTRICHUM CANDIDUM*
×2000

the entire surface; conidia 1-celled, globose, often variously colored in mass, catenulate, produced basipetally. A large genus containing many species saprophytic on a wide variety of substrata and a few parasitic species.

Typical conidial heads are shown in Fig. 4.7. Although differences are easily discerned between this *Aspergillus* and the *Penicillium* shown in Fig. 4.8, there are borderline species which are difficult to classify. Aspergilli have special cells called foot cells which the penicillia do not.

Courtesy of M. F. Brown and H. G. Brotzman, University
of Missouri-Columbia

FIG. 4.7.   CONIDIAL HEADS OF *ASPERGILLUS NIGER*

Genus *Penicillium*. Conidiophores arising from the mycelium singly or less often in synnemata, branched near the apex to form a brush-like, conidia-bearing apparatus; ending in phialides which pinch off conidia in dry chains; conidia hyaline or brightly colored in mass, 1-celled, mostly globose or ovoid, produced basipetally. A large genus containing both parasitic (producing rots of fleshy plant parts) and saprophytic species.

The broom-like whorl, giving the name to this genus, is evident in Fig. 4.8.

Genus *Cladosporium*. Conidiophores dark, branched variously near the apex or middle portions, clustered or single; conidia dark, 1- or 2-celled, variable in shape and size, ovoid to cylindrical and irregular, some typically lemon-shaped; parasitic on higher plants or saprophytic.

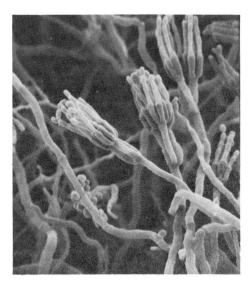

Courtesy of M. F. Brown and H. G. Brotzman, University of Missouri-Columbia

FIG. 4.8. THE PENICILLUS OF *PENICILLIUM ITALICUM*

SPECIFIC FOOD SPOILAGE FUNGI

Fungi which were discussed previously spoil specific foods. The most important groups are discussed below.

Class Oömycetes

Two important fungi in this class are *Phytophthora infestans* and *Pythium debaryanum*. *P. infestans* causes late blight of potatoes and tomatoes. *P. debaryanum* produces various colors on potatoes. Another species, *P. cactorum,* is responsible for leather rot in strawberries.

Class Zygomycetes

Rhizopus nigricans (*stolonifer*) causes spoilage of bread and cereal products, sweet potatoes and strawberries. *Thamnidium elegans* and *Mucor mucedo* produce whiskers on stored meats.

Class Ascomyetes

The only filamentous fungus we will study is *Monilinia* which causes brown rot of stone fruits.

Class Deuteromycetes (Fungi Imperfecti)

A great majority of spoilage fungi belong to the Deuteromycetes. The molds shown in Table 4.1 are not by any means a complete list.

FUNGI AS A SOURCE OF TOXINS

As research has progressed during the last few years, more and more mycotoxins have been described. At the present time, there are analytical methods for analysis of 4 mycotoxins in the Official Methods of Analysis (AOAC 1975). These toxins will be discussed in more detail later in the book under food poisoning.

TABLE 4.1.   SELECTED FUNGI IMPERFECTI THAT SPOIL FOODS

Fungi	Food Spoiled
Penicillium	
P. chrysogenum	Cheese and other dairy products, bread and pastries, fruits, vegetables, meat and meat products, improperly canned foods and decaying vegetation of all kinds.
P. digitatum	Causes green rot of citrus fruits. Fungus produces mummies.
P. italicum	Causes blue-green rot of citrus fruits (soft rot).
P. striatum	Causes spoilage of canned blueberries. Survives 93.3°C (200°F) and grows in cans with 12 to 15 in. of vacuum.
P. stoloniferum	Causes spoilage of cereal grains; spoils bread.
P. expansum	Spoils, grain, cereal products, eggs, chickens in storage. Soft rot of apples and grapes.
Aspergillus	
A. flavus-oryzae group	Spoils egg noodles, bread, bakery products, dried dates, cured meats, dairy products, nut meats.
A. glaucus group	Most common molds on earth. Spoil sweetened products such as jams, jellies, soft sugars, honey, soft candies, meats, pickles, dried foods.
A. versicolor group	Spoils breads, cereals, dried meats.
A. niger group	Spoils fruits, vegetables, dairy products.
Cladosporium sp.	Produces greenish spots on butter; causes green mold rot of fruits and vegetables.
Botrytis cinerea	Causes gray mold of fruits and vegetables.
Alternaria	
A. tenuis	Alternaria rot of vegetables.
A. solani	Causes early blight of tomatoes.
A. citri	Rots grapefruit.
Fusarium	Rots of fruit and vegetables.
Sporotrichum	Causes white spots on meats.

Source: Thom and Raper (1945); Raper and Thom (1949); Frazier (1967).

PRESERVATION OF MOLD CULTURES

Mold cultures can be kept on agar slants. Most of the cultures are transferred once a year. Some may require a transfer every 6 months. Cultures should be held in screw-top test tubes at 4°C.

A heavy mold spore suspension is used to inoculate sterile soil. After the soil is allowed to dry by screwing on the test tube cap loosely, it may be placed in refrigerated storage.

Another way of preservation is freeze-drying the culture and storing it under vacuum at 4°C. With this method, very high numbers are needed since there is some kill during the freezing and drying process.

Cultures of molds can also be kept under sterile mineral oil.

One must experiment with different strains since not all remain viable using the methods named above.

GLOSSARY

Ascospore—A spore, which results from meiosis, borne in an ascus.

Ascus—A sac-like structure generally containing a definite number of ascospores (typically eight); these are usually formed as a result of karyogamy and meiosis. Characteristic of the class Ascomycetes.

Budding—The production of a small outgrowth from a parent cell. A method of asexual reproduction.

Chlamydospore—A hyphal cell enveloped by a thick cell wall, which eventually becomes separated from the parent hypha and behaves as a resting spore.

Coenocytic—Non-septate; referring to the fact that nuclei are embedded in the cytoplasm without being separated by walls.

Conidium—A spore formed asexually, usually at the tip or side of a hypha.

Diploid—Containing the double (2N) number of chromosomes.

Fission—The splitting of a cell into two cells.

Fungus—One of the achlorophyllous thallophytes whose somatic structures are usually filamentous and branched. Fungi have cell walls and demonstrable nuclei. They reproduce typically by both sexual and asexual means.

Gametangium—A structure which contains gametes.

Gamete—A differentiated sex cell or a sex nucleus which fuses with another in sexual reproduction.

Haploid—Containing the reduced (N) number of chromosomes.

Heterothallic—Refers to a species in which the sexes are segregated into separate thalli; two different thalli are required for sexual reproduction.

Homothallic—Refers to fungi in which sexual reproduction takes place in a single thallus which is, therefore, essentially self-compatible.

Hyaline—Colorless, transparent.

Hypha—The unit of structure of the fungi; a tubular filament.

Karyogamy—The fusion of two nuclei.

Macroconidium—A conidium, as distinguished by a microconidium (as in *Fusarium*). Macroconidia of *Fusarium* are multicelled spores. Microconidia in *Fusarium* are small and 1-celled.

Meiosis—A pair of nuclear divisions in quick succession, one of which is reductional (from 2N to N).

Mycelium—Mass of hyphae constituting the body of a fungus.

Oidium—A thin-walled, free, hyphal cell derived from fragmentation of a somatic hypha into its component cells or from an oidiophore.

Oögonium—A female gametangium containing one or more eggs.

Oösphere—A large, naked, non-motile female gamete.

Oöspore—A thick-walled spore which develops from an oösphere either by fertilization or parthenogenesis.

Parthenogenesis—The development of the normal product of sexual reproduction from the female gamete alone.

Penicillus—The conidiophore of the genus *Penicillium*.

Phialide—A small, bottle-shaped structure in which spores are produced. The latter are characteristically formed inside the phialide and extruded.

Rhizomorph—A thick strand of somatic hyphae in which the hyphae have lost their individuality, the whole mass behaving as an organized unit.

Sclerotium—This hard resting body resistant to unfavorable conditions, may remain dormant for long periods of time and germinate upon return of a favorable environment.

Septate—With cross walls.

Soma—The body of an organism as distinguished from its reproductive organs or its reproductive phase.

Sporangiophore—A hypha which bears a sporangium.

Sporangium—A sac-like structure, the entire protoplasmic contents of which become converted into an indefinite number of spores.

Thallus—A relatively simple plant body devoid of stems, roots, and leaves; in fungi, the somatic phase.

Vesicle—The bulbous head terminating conidiophore of *Aspergillus*.

Zygospore—A resting spore which results from the fusion of two gametangia.

Zygote—A diploid cell resulting from the union of two haploid cells.

Zoospore—Motile spores produced sexually by the Oömycetes.

REVIEW QUESTIONS

1. What plant pathogen caused a famine in Ireland?
2. How serious is spoilage caused by filamentous fungi in destroying our food?
3. List some enzymes produced by filamentous fungi. What is one produced commercially?
4. Name molds used in the manufacture of cheeses.
5. List elements which can be assayed by filamentous fungi.
6. Discuss how to isolate molds.
7. How are filamentous fungi cultivated?
8. Discuss general characteristics of filamentous fungi.
9. List classes of fungi we are studying.
10. Discuss sexual reproduction in the following classes: Oömycetes, Zygomycetes and Ascomycetes.
11. Discuss asexual reproduction in Zygomycetes.
12. List nine genera in the Fungi Imperfecti which we are studying.
13. Make a list of fungi and the foods they spoil.
14. What kind of toxins do molds produce?
15. How can you preserve molds?
16. Know all terms in the glossary.

SUGGESTED LABORATORY EXERCISES

1. Examine cultures provided and make drawings.
2. Streak from skimmed milk cultures onto potato dextrose agar (PDA), and incubate at room temperature.
3. With an inoculating needle, mix a small amount of the fungus with some liquid PDA on a slide. Add a cover slip. Place in a sterile Petri plate and incubate at room temperature.
4. Study and draw all mold cultures.

REFERENCES

AOAC. 1975. Official Methods of Analysis of the Association of Official Analytical Chemists. Association of Official Analytical Chemists, Washington, D.C.

BARNETT, H. L. 1960. Illustrated Genera of Imperfect Fungi. Burgess Publishing Co., Minneapolis.

BBL. 1968. BBL Manual of Products and Laboratory Procedures, 5th Edition. BBL, Division of Becton, Dickinson and Co., Cockeysville, Maryland.

CHESTER, K. S. 1947. Nature and Prevention of Plant Diseases. The Blakeston Co., Philadelphia.

FRAZIER, W.C. 1967. Food Microbiology. McGraw-Hill Book Co., New York.

LILLY, V. G. and BARNETT, H. L. 1951. Physiology of the Fungi. McGraw-Hill Book Co., New York.

MOORE-LANDECKER, E. 1972. Fundamentals of the Fungi. Prentice-Hall, Engelwood Cliffs, N.J.

PRESCOTT, S. C. and DUNN, C. G. 1949. Industrial Microbiology. McGraw-Hill Book Co., New York.

RAPER, K. B. and THOM, C. 1949. A Manual of the Penicillia. Williams and Wilkins Co., Baltimore.

THOM, C. and RAPER, K. B. 1945. A Manual of the Aspergilli. Williams and Wilkins Co., Baltimore.

USDA. 1965. Losses in agriculture. Agric. Handbook *291*.

WOLF, F. A. and WOLF, F. T. 1947. The Fungi. John Wiley & Sons, New York.

5

Factors Influencing Growth of Spoilage Microorganisms

In this chapter we are going to consider various factors which influence the growth of microorganisms in foods and in nature as a whole. The spoilage flora varies with the chemical composition of foods and other substrates. In this chapter we will consider the following factors as influencing microbial growth: nitrogen and carbon sources, minerals, vitamins, water, osmotic pressure, oxidation-reduction potential, pH, antimicrobial substances, biological structure, and the relationship of one microorganism to another. Both foods and media will be considered.

NITROGEN AND CARBON SOURCES

Some microorganisms are able to use proteins as both a nitrogen and a carbon source. For example, proteolytic bacteria can grow in nutrient broth. In this medium, carbohydrates are absent or at low levels. The nutrient broth consists of beef extract and peptone. The ultimate source of nitrogen is the amino acid molecule. Bacteria have proteolytic enzymes which break down peptones to amino acids. The order of breakdown is proteins to proteoses to peptones to polypeptides to peptides and peptides to amino acids. The amino group is removed from the amino acid by deamination.

The process of decomposition of proteins to peptones is called peptonization; whereas the breakdown of proteins to amino acids is called proteolysis. Production of ammonia from amino acids is called ammonification. If the decomposition of proteins occurs in the presence of oxygen, the process is called decay; in the absence of oxygen, it is putrefaction. Whole proteins (such as casein) and proteoses are not attacked by most proteolytic bacteria unless a small amount of peptone is added.

Organic sources of nitrogen used by microorganisms include proteins (and their decomposition products), purines, amines, and amino acids. Urea, sodium nitrate, sodium nitrite and ammonium salts are inorganic sources. The ammonium ion (NH_4^+) is suitable for practically all yeasts. Some fungi can utilize ammonium nitrogen and are not able to assimilate nitrate nitrogen. Utilization of proteinaceous substances is used in identifications of both yeasts and bacteria.

Bacteria which spoil high protein foods (red meats, poultry and fish) are proteolytic and possibly many of the strains are also lipolytic. Although lipids are not required by microorganisms, their enzymes may produce rancidity.

Nitrogen is a part of all amino acids and hence a part of all proteins. Nitrogen is a part of the cytoplasm and cell wall and enzymes. Carbon is a part of the cytoplasm, enzymes, cell wall and reserve material in the cell. Microorganisms obtain energy by oxidizing carbon compounds. Monosaccharides (glucose, fructose, galactose), disaccharides (maltose, lactose, sucrose), trisaccharides (raffinose) and polysaccharides (starch, pectin and cellulose) are used in taxonomic studies of yeasts and bacteria. Derivatives of carbohydrates, such as organic acids (citric, malic, succinic, etc.), are also used in the same way. Glucose is utilized by more organisms than any other sugar. This is true for yeasts, molds, and bacteria.

MINERAL REQUIREMENTS

There are many elements which are needed in the nutrition of microorganisms. For example, the following metals are considered essential for fungi (Lilly and Barnett 1951): potassium, magnesium, iron, zinc, copper, calcium, gallium, manganese, molybdenum, cobalt, vanadium and scandium. Bacteria require potassium, calcium, magnesium, iron and manganese. Microorganisms also require nonmetallic elements like hydrogen, oxygen, and sulfur.

Manganese, magnesium and zinc are involved as coenzymes in glycolysis. Manganese, magnesium and iron act as coenzymes in the Krebs cycle.

HIGH PROTEIN FOODS AS A SOURCE OF N, C, AND MINERALS

The so-called high protein foods are relatively low in protein; of the four high protein foods listed in Table 5.1, tuna ranks first. Water makes up a major portion of each of these foods as does fat. Only in eggs is there any carbohydrate. Since microorganisms do not need fat, spoilage microor-

TABLE 5.1. COMPOSITION OF SELECTED HIGH PROTEIN FOODS

Food	Water (%)	Protein (%)	Fat (%)	Carbo-hydrate (%)	Ash (%)
Beef carcass (total edible, trimmed to retail level, raw)					
Choice Grade (75% lean, 25% fat)	56.7	17.4	25.1	0	0.8
Standard Grade (82% lean, 18% fat)	63.9	19.4	15.8	0	0.9
Chickens (all classes (raw))					
Light meat without skin	73.7	23.4	1.9	0	1.0
Dark meat without skin	73.7	20.6	4.7	0	1.0
Eggs (raw chicken)					
Whole	73.7	12.9	11.5	0.9	1.0
Whites	87.6	10.9	Tr	0.8	0.7
Yolks	51.1	16.0	30.6	0.6	1.7
Tuna (raw)					
Bluefin	70.5	25.2	4.1	0	1.3

Source: Watt and Merrill (1963).

ganisms obtain their N and C from the protein content. Minerals are found in the ash content.

VITAMINS

Microorganisms may be self-sufficient with respect to their need of vitamins provided in the growth medium. For example, all the molds which grow on Czapek's medium are self-sufficient. Since this medium contains no vitamins, a fungus synthesizes all the vitamins it needs. *Ashbya gossypii* and *Eremothecium ashbyii* may be used to produce riboflavin commercially (Moore-Landecker 1972). *Blakeslea trispora* produces beta-carotene, the precursor of vitamin A.

Some microorganisms may be partially or completely deficient in one or more vitamins. An example of a fungus completely deficient in thiamin is *Phycomyces blakesleeanus*. An example of a fungus being partially deficient in thiamin is *Lenzites trabea*. Both fungi require thiamin for growth.

Bacteria, molds and yeast may be used as test microorganisms for the microbiological assay of vitamins. Genera of microorganisms used to assay for vitamins are as follows (Freed 1966): *Lactobacillus, Leuconostoc, Acetobacter, Clostridium, Proteus, Streptococcus, Torula, Bacillus, Pediococcus, Euglena, Rhizobium, Shigella, Staphylococcus, Neurospora, Phycomyces, Endomyces, and Saccharomyces.*

Vitamins play a vital role in various metabolic functions of the cell. These functions are listed in Table 5.2.

TABLE 5.2.   METABOLIC FUNCTIONS OF THE VITAMIN B GROUP

Vitamin	Typical Organisms Requiring It	Coenzyme	Metabolic Function
Thiamin	Staph. aureus L. fermenti	DPT	Activation of keto acids and keto sugars; transfer of 2-carbon units
Nicotinic acid	L. arabinosus P. vulgaris	DPN and TPN	Hydrogen transfer
Riboflavin	L. casei S. lactis	Riboflavin-5-P Flavin-adenine- dinucleotide	Hydrogen transfer
Vitamin B$_6$ pyridoxal or pyridoxal or pyridoxamine	L. casei C. welchii S. fecalis	Pyridoxal-P Pyridoxamine-P	Decarboxylation, deamination, transamination and racemization of amino acids
Pantothenic acid	Brucella abortus P. morganii	CoA	Acyl activation and transfer
p-Aminobenzoic acid	C. acetobutylicum A. suboxydans	Tetrahydrofolic acid	1-Carbon transfer
Folic acid	L. casei C. tetani	Tetrahydrofolic acid	
Biotin	L. mesenteroides C. tetani L. arabinosus	Biotin-CO$_2$	CO$_2$ fixation, fatty acid synthesis
Vitamin B$_{12}$	L. leichmannii L. lactis	(?)	1-Carbon transfer synthesis of deoxyribosides

Source: Burrows et al. (1964).

WATER CONTENT, OSMOTIC PRESSURE AND WATER ACTIVITY

Water occurs in foods as free, bound and water of hydration. If the water content (free water) is lowered to 3–5% for vegetables and to 15–20% for fruits, microorganisms cannot grow. The reason that fruits can be dried to a higher moisture content than vegetables is because they have a higher natural sugar content before the food is dried.

Water Content

Water content of food products may be expressed as a percentage determined as follows:

$$\frac{\text{weight before drying} - \text{weight after drying}}{\text{weight before drying}} \times 100 = \% \text{ moisture}$$

Because foods contain different components such as starch, cellulose, protein, etc., the foods have differing water-holding capacities. The water content of raw figs is 77.5% (Watt and Merrill 1963), whereas the moisture content of endive is 93.1%.

Water Activity

The water activity of a food may be defined as the vapor pressure of the food divided by the vapor pressure of water at the same temperature. Experimentally, water activity (a_w) is determined in a closed container and the relative humidity above the food is measured. This determination is called equilibrium relative humidity (ERH). ERH divided by 100 gives a_w. On this scale, pure water vapor in a saturated atmosphere is equal to 1.0. Using this method of expressing moisture, one can divide the microflora into several groups. The lowest a_w values permitting growth are as follows (Mossel and Ingram 1955):

	a_w
Normal bacteria	0.91
Normal yeasts	0.88
Normal molds	0.80
Halophilic bacteria	0.75
Osmophilic yeasts	0.60

Osmotic Pressure

Osmotic pressure influences the growth of microorganisms. Osmotic pressure can be defined as the unbalanced pressure that gives rise to the phenomena of diffusion and osmosis, as when sucrose, contained in a collodion membrane, is placed in water. Osmotic pressure is proportional to the absolute temperature and also the molecular concentration. Osmotic pressure can be related to a_w by the following formula:

$$\text{Osmotic pressure} = \frac{-\text{RT} \log e \, a_w}{\overline{V}}$$

$\overline{V}$ is equal to the partial molal volume of water. Osmotic pressure is directly proportional to the absolute temperature (Scott 1957).

The relation between grams of sucrose per 100 g of water to a_w to molalities of sucrose is listed in Table 5.3.

Where do environments of high osmotic pressure occur in the food industry? Syrups (10–55%), commonly used in canning, and liquid sugar (67–72%), used in the food industry, are examples of high osmotic environments. Honey contains 70–80% sugar. High osmotic pressure environments or low a_w exists in dried foods and in brines. Brines are used in the fermentation of cucumbers to make "salt stock" for future pickle manufacture. At the start of the fermentation, a 10% sodium chloride solution is prepared. This has an a_w of 0.938. When the fermentation is finished, the percentage of salt is increased to 15% or a_w of 0.91. At this a_w and the acid content, the fermented cucumbers are preserved as long as film yeast does not grow on the surface of the vats and use the lactic acid.

TABLE 5.3. THE RELATION BETWEEN GRAMS OF SUCROSE PER 100 G WATER TO a_w TO MOLALITIES OF SUCROSE

g Sucrose to 100 g Water[1]	a_w at 25°C	Molalities of Sucrose at 25°C
9.3	0.995	0.272
18.3	0.990	0.534
35.3	0.980	1.03
65.7	0.960	1.92
93.1	0.940	2.72
119.1	0.920	3.48
140.7	0.900	4.11

Source: Scott (1957).
[1] Values added by this author.

Sorption Isotherms

Dried foods or wet foods may take up or give up moisture to the surrounding environment depending upon whether the surrounding air is drier or more moist. When the surrounding air has a relative humidity greater than ERH of the product, highly concentrated foods (dried or concentrates) draw moisture from the air and may thus become moist enough for spoilage microorganisms to grow.

A plot of ERH vs percentage of the moisture of the food is called a sorption isotherm. (Isotherm means that the product has the same temperature at a given time. Sorption means the taking up and holding of the moisture by either absorption or adsorption.) By constructing such curves, one can determine at what ERH a food product can be stored so that the product will not increase in moisture content.

Osmophilic Microorganisms

Osmophilic microorganisms grow well in high osmotic pressure environments (or low a_w environments). Strains of *Aspergillus glaucus, A. niger, Penicillium* sp. and *Saccharomyces rouxii* are examples of osmophilic flora which can grow in media having 40, 50, and 60% soluble solids.

Microorganisms capable of growing in high salt concentration may be called halophilic or osmophilic. Microorganisms growing on dry materials may be called either xerophilic (dry-loving) or osmophilic.

OXIDATION-REDUCTION POTENTIAL (Eh)

Definitions

Classification of bacteria may be based on their relationship to oxygen; for example, as aerobic, anaerobic or as facultative anaerobic. In an environment with oxygen, there can be a loss of electrons; in a reduced environment, there is a gain of electrons. In an actively growing culture, there may be an excess of electrons and, hence, a reduced condition may exist. This can be measured with dyes or electrodes. Methylene blue in the reduced state of the molecule is colorless and blue in the oxidized state.

Eh is measured in millivolts and indicates the oxidation-reduction state of the food or medium. Positive values indicate aerobic conditions, whereas negative values indicate an anaerobic environment. Various components in the food (such as reducing sugars, vitamin C, and sulfur-containing amino acids or other sulfur compounds having an –SH group) can influence the Eh of the food.

Eh Values of Foods

Mossel and Ingram (1955) listed foods that may be grouped as oxidized or reduced. These are shown in Table 5.4. By inspecting these data, we may observe that a minced liver medium has a low Eh. Hence, liver broth is an excellent medium for the growth of anaerobes.

Meat can support aerobic growth of bacteria in a slime on the surface while, at the same time, putrefactive anaerobic bacteria can grow deep within. In a liquid medium, aerobic bacteria can grow, lowering the Eh to a level at which the anaerobic bacteria start to grow and to lower the Eh further.

Table 5.5 lists examples of the lowest potentials reached in the medium after growth of two yeasts. By inspecting data in Table 5.5, one can see that the type of substrate and whether conditions are aerobic or anaerobic influence how low the Eh will drop.

If electrode potentials in the media are measured during the growth of haemolytic streptococci and *Corynebacterium diphtheriae,* initially in the medium containing the streptococci the Eh will be lowered and then the Eh will rise; whereas, the Eh of the growth medium of *C. diphtheriae* will remain at a low level (Hewitt 1950).

TABLE 5.4.   Eh VALUES OF OXIDIZED AND REDUCED FOOD GROUPS[1]

State of Food	Food	Eh(mv)
Oxidized	Pear juice	+436
	Grape juice	+409
	Lemon juice	+383
	Butter serum	+350 to +290
	Milk	+340 to +220
	Muscle, minced cooked (access to air)	+300
	Muscle, minced raw (access to air)	+225
Reduced	Cheese, Dutch	−20 to −310
	Cheese, Emmenthal	−50 to −200
	Muscle, post rigor raw	−150
	Liver, minced raw	−200
	Wheat (whole grain)	−320 to −360
	Wheat germ	−470

Source: Mossel and Ingram (1955).
[1] Data rearranged from most oxidized to most reduced.

TABLE 5.5.   LOWEST Eh VALUES REACHED IN MEDIA AFTER GROWTH OF YEASTS

Yeasts	Substrate	Condition	Eh(Mv)
Saccharomyces cerevisiae	lactose	aerobic	+310
S. cerevisiae	glucose	aerobic	+160
S. cerevisiae	glucose	anaerobic	+ 90
Torula candida	glucose	aerobic	+270

Source: Hewitt (1950).

The Role of SH Compounds in Lowering Eh in Media

Sodium thioglycollate ($NaOOCCH_2SH$) and L-cystine ($S\text{-}CH_2\text{-}CH(NH_2)COOH)_2$ are added to a fluid called thioglycollate medium. Sodium thioglycollate is in a reduced form; L-cystine is in the oxidized state; both compounds can undergo oxidation and reduction. The composition of this medium is as follows:

	g/liter
Pancreatic digest of casein	15
L-cystine	0.5
Dextrose	5.0
Yeast extract	5.0
Sodium chloride	2.5
Sodium thioglycollate	0.5
Resazurin (O-R dye)	0.001
Agar	0.75

The sodium thioglycollate (50 mg/100 ml) in this medium is added to lower the Eh so that anaerobes may be grown in a liquid medium. Electrode potentials of sodium thioglycollate added to distilled water are shown in Table 5.6.

The Eh of fluid thioglycollate was +292 or the medium had an Eobs of +50 mv. This is close to the value for 50 mg of sodium thioglycollate.

Another medium for the growth of anaerobes is Brewer's anaerobic agar. This medium was developed to be used in a special Petri plate so that anaerobes and microaerophiles can be grown on the surface of agar; thus the use of special equipment such as anaerobic jars was eliminated. The composition of this medium is as follows (Difco 1953):

	g/liter
Yeast extract	5
Tryptone	5
Proteose peptone	10
Dextrose	10
Sodium chloride	5
Agar	20
Sodium thioglycollate	2
Sodium formaldehyde sulfoxylate	1
Resazurin	0.002

TABLE 5.6.   EFFECT OF SODIUM THIOGLYCOLLATE ON Eh AT 20°C

Sodium Thioglycollate Added mg/100 ml	Eobs	+	E Cal	=	Eh
0	+210		+242	=	+452
50	+40		+242	=	+282
100	−60		+242	=	+182
150	−85		+242	=	+157
200	−100		+242	=	+142

Source: Fields (1976).

The observed electrode potential of Brewer's anaerobic medium after autoclaving was −340 mv or changed into Eh value of −98 (−340 + 242, for the potential of the calomel electrode at 20°C).

Another anaerobic agar for the detection of anaerobes producing H_2S is sulfite agar. This medium is composed of trypticase 10 g/liter, sodium sulfite 1 g/liter and agar 15 g/liter. The Eh of this medium in liquid state is +92.

Many of the clear liquid media which have been used for the cultivation of anaerobes contain SH groups for lowering the O-R potential. In addition to sodium thioglycollate, pyruvic acid, cysteine, and glutathione have been used. Dextrose is of value in promoting the growth of many bacteria and serves to prolong anaerobiosis. According to Brewer (1940), the fluid thioglycollate medium might be called a facultative medium since it supports the growth of most obligate clostridia as well as microaerophiles and aerobes.

pH

pH is an important factor influencing the growth of spoilage microorganisms. In Fig. 5.1, a food scientist is shown using a pH meter.

Definition.

$$\text{pH is equal to } \log \frac{1}{[H^+]} \text{ or } - \log [H^+]$$

pH Range of Foods. The pH of complex foods (eliminating vinegar) can range from 2.2 for lemons to 9.3 for old egg white. For the most part, however, the pH of foods will not be above a pH of 7.

pH Values Favorable for Bacteria. On the whole for optimum activity, bacteria like a more alkaline pH or especially a pH near to 7 for optimum activity.

Some examples of the range (4.0–9.6) and optimum pH for growth of bacteria are listed in Table 5.7.

Yeasts (2.4–9.2) and molds as a group tolerate a greater pH range than do bacteria. *Aspergillus oryzae* is reported to grow from pH 1.6 to 9.3 (Altman and Dittmer 1966).

The pH is a powerful force regulating microbial growth.

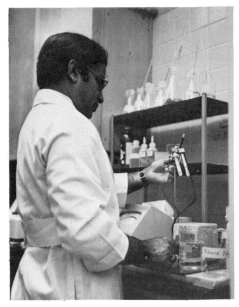

Courtesy of University of Missouri-Columbia

FIG. 5.1.   FOOD   SCIENTIST   USES   A   pH
METER

ANTIMICROBIAL SUBSTANCES FOUND IN FOODS

There are a few naturally-occurring compounds in foods which have antimicrobial activity. Lactinin occurs in milk, lysozyme in eggs and essential oils in spices. These will be covered in detail later in the book.

TABLE 5.7.   RANGE AND OPTIMUM pH FOR GROWTH OF SELECTED BACTERIA

Bacteria	Optimum pH	Range	Reference
Enterobacter aerogenes	6.0	4.4 to 9.0	(2)
Escherichia coli	6.0–7.0	4.3 to 9.5	(2)
Bacillus subtilis	6.0–7.5	4.5 to 8.5	(2)
Clostridium sporogenes	6.5–7.5	5.0 to 9.0	(2)
Erwinia carotovora		4.6 to 9.3	(2)
Flavobacterium aquatile	7.2–7.4	6.5 to 7.8	(1)
Micrococcus cryophilus	6.8–7.2	5.5 to 9.5	(1)
Proteus vulgaris	6.5	4.4 to 9.2	(2)
Pseudomonas aeruginosa	6.6–7.0	4.4 to 8.8	(2)
Salmonella typhi	6.8–7.2	4.0 to 9.6	(2)
Salmonella paratyphi	6.2–7.2	4.0 to 9.6	(2)
Shigella dysenteriae	7.0	4.5 to 9.6	(2)

[1] Breed *et al.* (1957).
[2] Thiamin (1963).

BIOLOGICAL STRUCTURE

There are various structures which protect food from spoilage microorganisms. For example, the shell protects the egg; the epidermis of vegetables and fruits protects them; the hide protects the flesh of animals. Healthy tissue in plants and animals is usually considered to be free of spoilage microorganisms. Spoilage microorganisms enter foods during processing and handling if disease-free plants and animals are used.

TEMPERATURE

Psychrophiles (−7° to 10°C)

These microorganisms (bacteria, molds) grow on refrigerated foods which are held for some time. They are not important as spoilage agents if the food is kept at room temperature. Psychrophilic bacteria spoil red meat, poultry meats, fish, shellfish, and dairy products, all of which are refrigerated.

Mesophiles (10° to 45°C)

Usually microorganisms with an optimum growth temperature of 35°–37°C are considered mesophilic. There is an overlapping with the upper part of the psychrophilic range and lower part of the thermophilic range. The mesophiles include the pathogenic microorganisms which attack man. This group also includes the food poisoning microorganisms which grow on all kinds of preserved and natural foods.

Thermophiles (30° to 73°C)

This group spoils canned foods by producing acid but no gas (flat sour). Thermophiles are found in soil, water, sugar, starch and spices. These bacteria produce very heat-resistant spores.

RELATION OF ONE MICROORGANISM TO ANOTHER

In a metabiotic association, one microorganism or microorganisms make conditions for the growth of the second organism or group. Usually, the microorganisms succeed each other and do not overlap too much in their growth patterns.

Metabiosis

Change in O-R Potential. An example of metabiosis occurs when aerobes grow and lower the Eh to a level at which anaerobes can grow.

Change in Acidity. Coliform bacteria exist at first during the fermentation of kraut but die off after production of acid. In a normal fermentation, *Leuconostoc mesenteroides* produces acid which enables *Lactobacillus plantarum* to multiply. *L. mesenteroides* dies out after the acid increases. *L. brevis* finishes the production of acid. Film yeasts may use the lactic acid and bring about spoilage of the kraut if it is held long enough before canning.

Change in Alcohol Content. Film yeasts can metabolize ethyl alcohol. If fermentative yeasts such as *Saccharomyces cerevisiae* convert the sugar into alcohol, film yeast may break down the alcohol into CO_2, water and yeast cells.

Competition of Food Poisoning Organisms with Spoilage Flora. Food poisoning bacteria may develop in very high numbers in pasteurized or heat-treated foods since there are no spoilage organisms to compete with them. If there are spoilage bacteria present in sufficiently high numbers, they produce off-odors which warn the consumer not to eat the food. This does not occur necessarily with food poisoning bacteria.

Color Changes in Foods. The growth of *Pseudomonas syncyanea* produces a light brownish tinge. When *Streptococcus lactis* grows in milk, there is no color change. However, if both grow together in milk, a bright blue color develops.

Antibiosis

Antibiosis is an antagonistic association between microorganisms. Certain strains of *Streptococcus lactis* produce nisin, an antibiotic. This, of course, can inhibit the growth of other bacteria.

Propionic acid produced by certain bacteria in Swiss cheese is inhibitory to molds. Alcohol formed in quantity by wine yeast inhibits its competitors. Acetic acid produced by some yeasts inhibits alcoholic fermentations.

GLOSSARY

Amines—Organic compounds that may be considered ammonia derivatives in which one or more hydrogen atoms have been replaced by a hydrocarbon radical.

Antibiotic—Substances produced by fungi, bacteria, and other organisms that inhibit or destroy other microorganisms.

C source—A carbon source is usually a sugar, although some microorganisms can obtain both carbon and nitrogen from amino acids or a single amino acid.

Coenzyme—A heat-stable organic molecule that must be loosely associated with an enzyme for it to function.

Czapek medium—A medium used to identify aspergilli and penicillia. The ingredients are all inorganic except the sugar, sucrose.

Equilibrium relative humidity—The relative humidity above a substance in a closed container.

Fermentation—Chemical reactions, induced by living organisms, which split complex organic molecules into relatively simple substances.

Film yeasts—Yeasts which form a film or pellicle on the surface of the growth medium.

Halophilic—Salt-loving.

Lipolytic—Splitting of fats.

Mesophile—A microorganism that usually has an optimum growth temperature at 35°–37°C.

N source—The nitrogen source in a growth medium may be either an inorganic salt containing N or a protein. Some simple media may have a single amino acid as the N source.

Osmotic pressure—Pressure created by different concentrations of substances (i.e., sugar and water) on either side of a semipermeable membrane.

Oxidation-reduction potential (Eh)—The degree of being oxidized or reduced. The potential is measured in volts or millivolts. An oxidized state will have a positive potential. A reduced state will have a negative potential.

pH—pH is the measure of acidity on a scale of 0 to 14. pH 7 is neutral. pH values less than 7 are acidic; above 7 are alkaline.

$$pH = \log 1/[H^+]$$

Psychrophiles—Microorganisms that grow at a low temperature, usually considered to be 15°C and below.

Purines—Organic nitrogenous compounds derived from or having molecular structures related to uric acid, adenine and guanine. These compounds are needed for synthesis of nucleic acids which make up RNA and DNA.

Sorption isotherm—Sorption means the process of sorbing or to take up and hold. Isotherm means a line drawn on a graph linking all points having identical mean temperature for a given period.

Thermophiles—Microorganisms that grow at high temperatures. Their optimum temperature is 55°C.

Urea—This compound, found in urine, is synthesized from ammonia and carbon dioxide. It is used in fertilizer and animal feed.

Water activity—The amount of water available is expressed on a scale of 1.0 to 0.0. Pure water is 1.0 on the a_w scale. An a_w of 0.7 is one at which most microorganisms can grow.

REVIEW QUESTIONS

1. Why do microorganisms need a source of N and C?
2. Discuss the breakdown of proteins.
3. Why does a microorganism need to break down proteins?
4. List compounds that furnish N to microorganisms.
5. Are high-protein foods really high in proteins?
6. Do microorganisms need fats?
7. Why do microorganisms need minerals?
8. Why do microorganisms need vitamins?
9. What role do vitamins play in the metabolism of bacteria?
10. Discuss the relationship between water content, osmotic pressure and a_w.
11. List the lowest a_w allowing growth for normal bacteria, normal molds, normal yeasts, osmophilic yeasts, and halophilic bacteria.
12. List foods or environments of high osmotic pressure in the food field.
13. Of what good are sorption isotherms?
14. Define Eh.
15. List some foods with low Eh values.
16. Does substrate and/or aerobic or anaerobic conditions have anything to do with the Eh of a growth medium after the growth of yeasts?
17. What compounds are added to media to lower the Eh?
18. Eh is measured in what units?
19. How is pH defined?
20. What is the range of pH of all foods? pH values are from 0 to 14.
21. Do most foods have a narrow or broad pH range?
22. What is the range of pH in which bacteria, yeast and molds grow?
23. What would you call a bacterium which loves to grow in an acid environment?
24. Define three groups of microorganisms that grow at different temperatures.
25. Give examples of antibiosis and metabiosis.

REFERENCES

ALTMAN, P. L. and DITTMER, D. S. (Editors). 1966. Environmental Biology. Federation of American Societies for Experimental Biology, Bethesda, Maryland.

BREED, R. S., MURRAY, E. G. D. and SMITH, N. R. (Editors). 1957. Bergey's Manual of Determinative Bacteriology. Williams and Wilkins Co., Baltimore.

BREWER, J. 1940. Clear liquid mediums for the aerobic cultivation of anaerobes. J. Am. Med. Assoc. *115*, 598–600.

BURROWS, W., MOULDER, J. W., and LEWERT, R. M. 1964. Textbook in Microbiology. W. B. Saunders Co., Philadelphia.

DIFCO LABORATORIES. 1953. Difco Manual of Dehydrated Culture Media and Reagents for Microbiological and Chemical Laboratory Procedures, 9th Edition. Difco Laboratories, Detroit.

FIELDS, M. L. 1976. The influence of different quantities of sodium thioglycollate on the Eh. Unpublished data. University of Missouri, Columbia.

FREED, M. 1966. Methods of Vitamin Assay. (Association Vitamin Chemists), Interscience Publishers, New York.

HEWITT, L. F. 1950. Oxidation-Reduction Potentials in Bacteriology and Biochemistry. E & S Livingstone, Edinburgh, Scotland.

LILLY, V. G. and BARNETT, H. L. 1951. Physiology of the Fungi. McGraw-Hill Book Co., New York.

MOORE-LANDECKER, E. 1972. Fundamentals of the Fungi. Prentice-Hall, Englewood Cliffs, N.J.

MOSSEL, D. A. A. and INGRAM, M. 1955. The physiology of the microbial spoilage of foods. J. Appl. Bacteriol. *18*, 232–268.

SCOTT, W. J. 1957. Water relations of food spoilage microorganisms. Adv. Food Res. *7*, 83–127.

THIAMIN, K. V. 1963. The Life of Bacteria. Macmillan Co., New York.

WATT, B. K. and MERRILL, A. L. 1963. Composition of Foods. Agric. Handbook *8*, USDA, Washington, D.C.

Microbiology of Fresh Fruits and Vegetables

SPOILAGE OF VEGETABLES BY NONPATHOGENIC BACTERIA

The Beijerinck principle states that deterioration of a food (all kinds, not just fruits and vegetables) is actually caused by only a small proportion of the microflora initially present so that a specific type of spoilage develops in a food under normal conditions of storage. These normal or standard conditions require the maintenance of appropriate temperatures. Also it means that there are no plant pathogens to attack the food during this time.

Mossel and Ingram (1955) suggested that factors which influence the genesis of association of the dominant flora determine the type of spoilage. These factors include (1) the initial infection, (2) the properties of the substrate, (3) conditions of storage, and (4) properties of the dominant microorganisms. Gram-negative rods (*Pseudomonas* and *Flavobacterium*), *Lactobacillus* and *Bacillus* are found as the dominant flora when spoilage occurs during standard conditions of storage. These soil bacteria get on the produce while they are being grown. These organisms may be called the autochthonous or indigenous flora.

SPOILAGE OF VEGETABLES BY PATHOGENIC FUNGI AND BACTERIA

Pathogenic bacteria and filamentous fungi which infect vegetables and cause rots are listed in Table 6.1.

TABLE 6.1. DISEASES OF VEGETABLES AND RECOMMENDED TEMPERATURES OF STORAGE

Vegetable	Disease	Control
Artichoke	Gray mold rot	Refrigerate at 0°C (32°F); and store in dry place
Asparagus	Gray mold rot	Refrigerate at 0°C (32°F);
	Fusarium rot	handle promptly
	Phytophthora rot	
	Bacterial soft rot	
	Watery soft rot	
Beans	Anthracnose	Refrigerate at 7.22°C (45°F)
	Bacterial soft rot	
	Rhizopus soft rot	
Cabbage	Bacterial soft rot	Store at 0°C (32°F)
Carrots	Bacterial soft rot	Avoid injury and refrig-
	Gray mold rot	erate at 0°C (32°F)
	Rhizopus soft rot	
	Watery soft rot	
Cauliflower	Bacterial soft rot	Refrigerate at 0°C (32°F)
Celery	Bacterial soft rot	
	Watery soft rot	Refrigerate at 0°C (32°F)
Cucumbers	Bacterial soft rot	Refrigerate at 7.22°–10°C (45°–50°F)
Eggplant	Bacterial soft rot	Refrigerate at 7.22°–10°C
	Alternaria rot	(45°–50°F)
Lettuce	Bacterial soft rot	Refrigerate at 0°C (32°F)
	Watery soft rot	
Muskmelons	Alternaria rot	Refrigerate 7.22°–10°C (40°–45°F)
	Cladosporium rot	during transit; 0°–1.1°C (32°–
	Fusarium rot	34°F) during storage;
	Phytophthora rot	avoid injuries
	Rhizopus rot	
Onions	Bacterial soft rot	Refrigerate at 0°C (32°F)
	Fusarium rot	
	Gray mold rot	
Parsley	Bacterial soft rot	Refrigerate at 0°C (32°F)
	Watery soft rot	
Parsnips	Bacterial soft rot	Refrigerate at 0°C (32°F)
	Gray mold rot	
	Watery soft rot	
Peas	Bacterial soft rot	Refrigerate at 0°C (32°F)
	Gray mold rot	
	Watery soft rot	
Peppers	Bacterial soft rot	Refrigerate at 7.22°C (45°F)
Potatoes	Bacterial soft rot	Refrigerate at 4.4°–10°C
	Fusarium rot	(40°–50°F)
	Bacterial ring rot	
	Late blight	

TABLE 6.1.   (*Continued*)

Vegetable	Disease	Control
Rutabagas	Gray mold rot	Refrigerate at 0°C (32°F)
Spinach	Bacterial soft rot	Refrigerate at 0°C (32°F)
Squash (winter)	Rhizopus soft rot	Refrigerate at 10°–12.7°C (50°–55°F)
Sweet potatoes	Rhizopus soft rot	Refrigerate at 12.7°–15.5°C (55°–60°F)
Tomatoes	Bacterial soft rot Alternaria rot Anthracnose rot Buckeye rot Gray mold rot Fusarium rot Watery rot	Refrigerate at 10°C (50°F)

Source: Chester (1947); Ramsey *et al.* (1952); Smith *et al.* (1966); Wright *et al.* (1954).

In addition to refrigeration, the following measures help to control spoilage: (1) eliminate all contaminated vegetables prior to storage; (2) use a chlorine wash to reduce the flora on surface; (3) dry excess moisture from wash; (4) hydrocool the vegetables rapidly to an acceptable temperature; (5) wax cucumbers and all root crops except Irish potatoes; (6) handle fruits and vegetables carefully to prevent breaks in the skin.

Inspection of the rots which occur on vegetables (Table 6.1) indicates that many of the types are repeated. Listed in Table 6.2 are the types of rots with the generic names of the causative microorganisms.

The following discussion will include some of the microbial species in Tables 6.1 and 6.2 as well as other causative bacteria.

TABLE 6.2.   MICROORGANISMS CAUSING DISEASES OF VEGETABLES

Disease	Microorganism
Gray mold rot	*Botrytis cinerea*
Phytophthora rot	*Phytophthora* sp.
Fusarium rot	*Fusarium* sp.
Bacterial soft rot	*Erwinia carotovora*
Watery soft rot	*Sclerotinia sclerotiorum*
Watery rot	*Oospora lactis* var. *parasitica*
Alternaria rot	*Alternaria* sp.
Cladosporium rot	*Cladosporium* sp.
Anthracnose	*Colletotrichum* sp.
Buckeye rot	*Phytophthora* sp.
Late blight	*Phytophthora infestans*

Source: Chester (1947); Ramsey *et al.* (1952); Smith *et al.* (1966).

Bacterial Spoilage

Bean blight (*Xanthomonas phaseoli*) and halo blight on beans (*Pseudomonas phaseolicola*) first show small water-soaked areas on pods which enlarge, dry out and become brick red in color. (Distinctions between the above diseases are observed on the leaves, not on the pods.)

Bacterial Canker, Spot and Speck of Tomatoes

1.   Bacterial canker (*Corynebacterium michiganense*) causes destruction of stems, leaves and spotting or decaying the fruit.

2.   Bacterial spot (*Xanthomonas vesicatoria*) produces small, black, scabby, roundish fruit spots, sometimes with water-soaked border.

3.   Bacterial speck (*Pseudomonas syringae*) produces small, numerous, dark brown, raised fruit spots, less than $\frac{1}{16}$ of an inch in diameter. This influences the appearance of the fruit.

Bacterial Soft Rot of Vegetables (*Erwinia carotovora*)

Erwinia carotovora produces a soft, foul-smelling, wet decay of root crops, crucifers, cucurbits, solanaceous vegetables, onions and flower bulbs. Soft rot bacteria cause spoilage of these crops during storage. This bacterium may be very destructive to cabbage, potatoes and other crops in the field since the source of these bacteria is the soil.

Proper maturing and drying of crops prior to storage will minimize losses. Since these organisms, which enter the vegetables through wounds, are more destructive at high temperatures, storage in a cool, dry place in well-aerated thin layers and avoidance of unnecessary bruising help to prevent spoilage (Chester 1947).

Bacterial Ring Rot of Potato (*Corynebacterium sepedonicum*)

The vascular ring, when invaded by bacteria, becomes creamy yellow or light brown and crumbly or cheesy without odor. This rot is hard to distinguish from *Fusarium* infection of potato except by microscopic and cultural methods.

Common Scab of Potato (*Streptomyces scabies*)

Small, brownish spots are formed on tubers which later become enlarged to form hard, circular or irregular corky areas. The spots may be cracked open. However, the cracks extend for only a little distance down into the flesh of the tuber. Scab lesions may form rough areas that may involve most of the tuber surface. The scabs are due to abnormal production of cork cells.

TABLE 6.3.   SPOILAGE DURING MARKETING OF VEGETABLES

Vegetable	Affected with Decay per Car (%)	Type of Spoilage
Artichokes	0–8	Mostly gray mold
Asparagus	0–44	Bacterial soft rot, phytophthora and blue mold
Beans, lima	1	Gray mold, bacterial soft rot, bacterial blight, watery soft rot
Beans, snap	8	Watery soft rot, soil rot, pythium rot, and gray mold
Broccoli	0–10	Bacterial soft rot and gray mold rot
Brussels sprouts	20–39	Bacterial soft rot
Cabbage	0–26	Bacterial soft rot, watery soft rot, gray mold rot, black rot and gray mold
Carrots	0–50	Bacterial soft rot, watery soft rot
Cauliflower	3	Bacterial soft rot, watery soft rot, and alternaria rot of the curd
Celery	Tr–50	Watery soft rot, bacterial soft rot
Cucumber	Tr–24	Bacterial soft rot, watery soft rot
Lettuce	0–50	Bacterial soft rot and watery rot
Onions	0–50	Bacterial soft rot, gray mold rot, black mold rot and smudge
Peppers	0–50	Rhizopus rot, bacterial soft, gray mold rot and alternaria rot
Potatoes	0–50	Bacterial soft rot (mostly)
Spinach	0–8	Bacterial soft rot
Sweet potato	1	Rhizopus and fusarium rot
Tomato	0–50	Bacterial soft rot, rhizopus rot, alternaria, watery rot, soil rot, gray mold rot, late blight rot, etc.

Source: USDA (1965).
[1] No data given.

Late Blight of Potato and Tomato (*Phytophthora infestans*)

Dr. Bellingham of Dublin, 1844 (Chester 1947) made this statement:

Toward the close of the month of August, I observed the leaves to be marked with black spots, as if ink had been sprinkled over them. They began to wither, emitting an offensive odor; and before a fortnight, the field, which had been singularly luxuriant and almost rank, became arid and dried up, as if by a severe frost. I had the potatoes dry out during the month of September, when about two-thirds were either positively rotten, partly decayed and swarming with worms, or spotted with brownish colored patches resembling flesh that had been frostbitten. These parts were soft to the touch and upon the decayed potatoes I observed a whitish substance like mold.

The tuber lesions resemble shallow pits which deepen when secondary decay sets in.

On tomatoes, the fruits are attacked when green, with large poorly defined brownish lesions involving up to half the fruit or more. The rot is firm, ordinarily not deep-seated.

Sweet Potato Rot and Strawberry Leak (*Rhizopus stolonifer*)

In sweet potato, the principal symptom is a rapid soft decay. The internal tissues become mushy and stringy, and the watery exudate often wets adjacent roots.

The signs of *Rhizopus* decay include the odor of fermentation, described in sweet potato as being first yeast-like, later resembling the odor of wild roses or geranium. In apples and peaches and in the later stages of decay in sweet potato, the tissues are darkly discolored; this does not occur in strawberries and grapes.

Early Blight of Potato and Tomato (*Alternaria solani*)

Alternaria causes a hard, black rot on the fruit, usually occurring at the stem end.

In the case of the potato, circular, decayed lesions may form on the tubers, permitting the entrance of secondary infections.

All common commercial varieties of potatoes and tomatoes are susceptible to early blight.

Tomato Anthracnose (*Colletotrichum phomoides*)

On tomato fruits, this mold commonly forms spots that at first are water-soaked. Later these areas are dark, depressed target-like rings which in moist weather are covered with masses of salmon-colored spores. Infection with *C. phomoides* results in exceedingly high mold counts in tomato products.

The influence of the microorganisms on actual losses of vegetables as they passed through marketing channels is given in Table 6.3. Even in modern times, filamentous fungi and bacteria take a great toll of our foods. Today with the population increase, it is extremely vital to increase the available food by decreasing spoilage to the lowest level possible.

SPOILAGE OF FRUITS BY PATHOGENIC FUNGI

Fruits are also spoiled as they pass through marketing channels. Examples of losses of fruits are listed in Table 6.4. As in vegetables, the same types of spoilage affect many fruits.

A life cycle of an ascomycete, *Monilinia fructicola,* is given in Fig. 6.1. Both asexual and sexual spores (ascospores) are involved in the life cycle. Ascospores are responsible for infection of blooms of the peach tree in the

TABLE 6.4.   SPOILAGE DURING MARKETING OF FRUITS

Fruit	Affected with Decay per Car (%)	Type of Spoilage
Apples	2	Blue mold, bull's eye rot
Apricots	0–19	Brown rot, rhizopus and gray mold
Cherries	0–49	Rhizopus, green mold and brown rot and gray mold
Grapefruit	0–16	Blue mold and stem end rot
Grapes	1	Gray mold and rhizopus
Lemons	0–8[1]	Blue mold, brown rot and alternaria rot
Limes	5	Blue mold
Oranges	0–25	Blue mold and stem end rot
Peaches	0–35	Mostly brown rot and rhizopus rot
Pears	0–25	Blue and gray mold
Plums and prunes	0–11	Rhizopus rot and blue mold
Strawberries	1–50	Gray mold and rhizopus rot
tangerines	0–15	Blue mold and stem end rot

Source: USDA (1965).
[1] Transit and unloading losses.

spring. Conidia or asexual spores produced on infected fruit spread the fungus from fruit to fruit.

SPOILAGE OF FRUIT JUICES BY NONPATHOGENIC MICROORGANISMS

In this section, we again return to the dominant spoilage genera as listed by Mossel and Ingram (1955). Under standard conditions of storage, two bacteria (*Acetobacter* and *Lactobacillus*), two yeasts (*Saccharomyces* and *Torulopsis*) and three filamentous fungi (*Botrytis*, *Penicillium* and *Rhizopus*) predominate. Note that the only bacteria named are acidophilic.

DETECTION OF FILAMENTOUS FUNGI IN FRUIT AND VEGETABLE PRODUCTS

The Howard Mold Count Method is used to determine whether or not the following food products have a high number of mold hyphae per microscopic field: butter, canned citrus and pineapple juices, cranberry sauces, frozen strawberries, puréed infant foods and tomato products.

In evaluating tomato products, a well-mixed sample is placed on a special slide so that the material examined will have a definite volume. At least 25 separate microscopic fields are examined per slide. If three pieces of hyphae

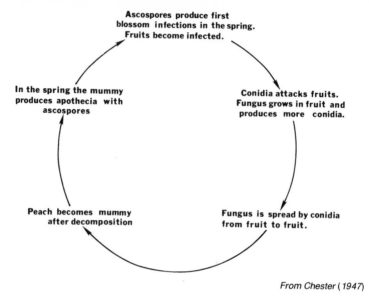

FIG. 6.1.   LIFE CYCLE OF *MONILINIA FRUCTICOLA*

measure $\frac{1}{6}$ the diameter of the field, the field is positive. If it takes more than three pieces of hyphae to make $\frac{1}{6}$ the diameter, the field is negative. If there is an excess amount of mold filaments, poor sorting of the raw product is indicated. There has been a certain tolerance in tomato products permitted over the years. For example, a mold count of 20% for tomato juice and 40% for catsup and purée has been tolerated. Counts above this value indicate that the product contained decomposed material. The 20% and 40% refer to the percentages of microscopic fields positive for mold hyphae; the product does not contain 20% or 40% rot by weight.

DETECTION OF DECOMPOSITION OF FRUIT AND VEGETABLE PRODUCTS BY CHEMICAL METHODS

Fields *et al.* (1968) discussed the use of metabolic by-products to determine food quality. Metabolic by-products associated with fruit and vegetable products are:

1.   Acetylmethylcarbinol (AMC).
2.   Diacetyl.
3.   2,3-butanediol (2,3-butylene glycol).
4.   Ethyl alcohol.
5.   Volatile and nonvolatile acids.

Characteristics of a good chemical indicator are as follows:

1. Nondecomposed food should not contain the indicator compound.
2. Indicator should increase with increased decomposition.
3. The indicator should be stable during processing and storage.

A standard of quality for apple juice was suggested by Fields (1962):

1. 0–1.0 ppm AMC indicates good or acceptable quality.
2. 1.0–1.6 ppm AMC indicate questionable quality.
3. Values greater than 1.6 indicate objectionable quality.

PUBLIC HEALTH ASPECTS OF FRESH FRUITS AND VEGETABLES

Human feces should not be used to fertilize fruits and vegetables. Pathogenic bacteria and amoebas in feces may survive for a while in soil and may contaminate fruits and vegetables, many of which are eaten raw. Washing and disinfecting of these foods can eliminate some of the problems. A 0.2% solution of chloride of lime used for 30 min destroys *Escherichia coli* on strawberries, lettuce and carrots.

GLOSSARY

Anthracnose—Any of several diseases of plants caused by fungi, in our case by *Colletotrichum* spp.
Autochthonous—Native to a particular place; indigenous.
Bacteria soft rot—Storage rot caused by *Erwinia carotovora.*
Beijerinck's principle—Beijerinck, a Dutch botanist, stated that only a small number of microorganisms actually cause the spoilage, although many kinds may be present in the spoiled food as secondary invaders.
Blight—Plant diseases that result in sudden dying of leaves, growing tips, or an entire plant. We list two blights in this chapter. See early and late blights.
Blue mold—Refers to penicillia.
Brown rot—A rot of stone fruits caused by *Monilinia fructicola.*
Bull's eye rot—Rot caused bv *Phytophthora* sp. on tomato fruits.
Canker—A necrotic area in a plant surrounded by healthy tissue.

Crucifers—Plants of the family Cruciferae. These plants have four-petaled flowers suggestive of a cross. Mustard is in this family.

Cucurbits—Plants that belong to the family Cucurbitaceae (squash, pumpkin and cucumbers).

Dominant flora—The dominant flora are those microorganisms that cause the spoilage under standard storage and handling conditions. (This is more or less a restatement of Beijerinck's principle.)

Downy mildew—A plant disease caused by fungi that belong to the order Peronosporales. The lower surface of the leaves has gray or velvety patches of spores.

Early blight—A plant disease of potato and tomato caused by *Alternaria solani*.

Gray mold—Gray mold is a common name to describe *Botrytis cinerea*.

Green mold—Either penicillia or aspergilli.

Howard mold count method—A method to determine mold hyphae in food products.

Indigenous flora—Flora native to a particular place. See autochthonous.

Late blight—Late blight is a disease caused by *Phytophthora infestans* on tomato and potato.

Leak—Soft rot allowing the contents to escape as in strawberry leak.

Phoma rot—Rot caused by *Phoma* sp.

Scald—A superficial discoloration on fruit, vegetables and leaves.

Smudge—Disease in which the fungus makes the vegetable appear dirty or sooty.

Solanaceous vegetables—Vegetables belonging to the family Solanaceae.

Watery soft rot—Rot caused by *Oospora lactis* var. *parasitica* on tomato fruits.

White rust—Fungal disease of crucifers caused by *Albugo candida*. This disease affects cabbage, cauliflower, cress, mustard, horseradish, rutabaga, and turnip.

REVIEW QUESTIONS

1. If human feces were used to fertilize vegetables and strawberries, what diseases might be transmitted by these foods to the consumer?
2. What was Beijerinck's principle?
3. What are the factors which influence the genesis of association of the dominant flora which spoil foods?
4. List the dominant flora of bacteria (nonpathogenic) which spoil vegetables.
5. List control factors to prevent spoilage of vegetables and fruits.

6. List microorganisms which cause rots in vegetables.
7. What is the most important pathogenic bacterium causing rots in vegetables?
8. What kind of enzymes produced by microorganisms causes soft rots?
9. What famine did *Phytophthora infestans* cause?
10. How serious is spoilage of fruits and vegetables caused by microorganisms as these foods proceed through marketing channels?
11. Describe the life cycle of *Monilinia fructicola*.
12. What are the dominant flora causing spoilage of fruit juices?
13. What is the Howard Mold Count Method?
14. What metabolic by-products are used to detect decomposition in vegetable products?
15. What chemical compound is used as a chemical indicator of quality of apple juice?
16. What are the public health aspects of eating raw fruits and vegetables?

REFERENCES

CHESTER, K. S. 1947. Nature and Prevention of Plant Diseases. The Blackston Co., Philadelphia.

FIELDS, M. L. 1962. Voges-Proskauer test as a chemical index to the microbial quality of apple juice. Food Technol. *16,* 98–100.

FIELDS, M. L., RICHMOND, B. S. and BALDWIN, R. E. 1968. Food quality as determined by metabolic by-products of microorganisms. Adv. Food Res. *16,* 161–229.

MOSSEL, D. A. A. and INGRAM, M. 1955. The physiology of the microbial spoilage of foods. J. Appl. Bacteriol. *18,* 232–268.

RAMSEY, G. B., WIANT, J. S. and MCCOLLOCK, L. P. 1952. Market diseases of tomatoes, peppers and egg plants. USDA Agric. Handbook *28.*

SMITH, M. A., MCCOLLOCH, L. P. and FRIEDMAN, B. A. 1966. Market diseases of asparagus, onions, beans, peas, carrots, celery and related vegetables. USDA Agric. Handbook *303.*

USDA. 1965. Losses in agriculture. Agric. Handbook *291.*

WRIGHT, R. C., ROSE, D. H. and WHITEMAN, T. M. 1954. The commercial storage of fruits, vegetables and florists and nursery stocks. USDA Agric. Handbook *66.*

7

Microbiology of Red Meats

PROCESSING (SLAUGHTERING), GENERAL CONDITIONS

Antemortem Conditions of Animals

The condition of the animals prior to slaughter is very important. One of the most important factors influencing the growth of microorganisms is pH of the substrate. After an animal is killed, there is conversion of glycogen in the muscle to lactic acid. The pH of the living animal is approximately 7. Within 6–8 hr postmortem, the pH decreases to 5.6–5.7 in pork muscle and finally, within 24 hr, the pH drops to 5.3–5.7 (Forrest *et al.* 1975). This occurs, of course, if the animal is rested and well fed. Animals under stress and excited will use the stored glycogen. Therefore, after slaughter the pH of the animal tissue will remain near 7 which permits rapid attack by spoilage microorganisms.

Data listed in Table 7.1 suggest that beef and pork are excellent substrates for microbial growth. Since beef and pork do not contain measurable carbohydrates, it is obvious that the microflora attacking beef and pork are highly proteolytic. They obtain both their N and C from amino acids. Meat is a good source of vitamins and also contains minerals. As mentioned previously, microorganisms influence fats which are not needed for their growth.

Sources of Contamination

As indicated in Table 7.2, there are many sources of contamination, even in very sanitary processing plants. It is highly important, therefore, to maintain the plant and the equipment in a clean and sanitary manner. Workers should wear clean clothes and have good work habits. These

TABLE 7.1.   FACTORS INFLUENCING MICROBIAL GROWTH ON BEEF AND PORK CARCASSES

	Beef[1] Carcass	Pork[2] Carcass
Nitrogen sources as protein	18.0%	7.9%
Carbon source as carbohydrate	0.0%	0.0%
Ash (minerals)	0.9%	[3]
Vitamins		
Thiamin	0.08 mg/100 g	0.39 mg/100 g
Riboflavin	0.16 mg/100 g	0.09 mg/100 g
Niacin	4.3 mg/100 g	2.1 mg/100 g
Water	60.1%	(leg 54%)[3,4]
Fat	21.0%	40.8%
pH	5.5–6.8[4]	5.3–5.7[5]
O-R	Aerobic on surface and anaerobic deep in tissue	Aerobic on surface and anaerobic deep in tissue

[1] Standard (73% lean, 27% fat) (Watt and Merrill 1963).
[2] Medium fat class (37% lean, 42% fat) (Watt and Merrill 1963).
[3] Not listed in USDA Handbook 8.
[4] According to Weiser (1971) for dried beef.
[5] According to Forrest et al. (1975).

principles are applicable not only to handling meat at slaughterhouses but also to wholesale distributors and retail grocery stores. Naumann et al. (1964) listed a cleaning schedule for a meat department in a retail store (Table 7.3). Cleaners and sanitizers should be used daily when slaughtering in an abattoir.

Romans and Ziegler (1974) pointed out that there are seven basic requirements in all federally inspected plants. These are (1) ante- and postmortem inspection; (2) reinspection after processing; (3) sanitation; (4) potable water; (5) sewage and waste disposal control; (6) pest control; and (7) condemned and inedible material control. If these are carried out according to federal requirements, most of the problems will be solved concerning sanitation. Good management must be exercised on a day-to-day basis.

TABLE 7.2.   SOURCES OF MICROBIAL CONTAMINATION DURING SLAUGHTERING OF BEEF AND PORK CARCASSES

From Animal Being Slaughtered	Vermin	Workers and Practices	Equipment	Building
Hides	Rats	Hands	Saws	Floor
Hoofs	Mice	Clothes	Knives	Walls
Hair	Insects	Quality	Skinning	Ceiling
Soil		of water	racks	
Feces		Sawdust	Scald vat	
Viscera		on floors	Conveyors	
			Table tops	
			Meat blocks	

TABLE 7.3.   MEAT DEPARTMENT CLEANING SCHEDULE

	Frequency	Scrape	Hot Water Rinse	Scrub with Cleaner	Hot Water Rinse	Sanitize	Squeegee	Air Dry
Equipment								
Band saw, grinder, cuber, etc.	Daily	X	X	X	X	X	X	X
Knives, scrapers, hand saws, etc.	Daily	X	X	X	X	X		X
Used trolley hooks, carts, etc.	Daily		X	X	X	X		X
Pans and lugs	Daily	X	X	X	X	X		X
Cutting blocks and tables	Daily	X	X	X	X	X		X
Floors								
Cooler	Weekly	X	X	X	X	X	X	X
Processing room	Daily	X	X	X	X	X	X	X
Walls								
Cooler	Weekly		X	X	X	X	X	X
Processing room	Weekly		X	X	X	X		X
Display case								
Outside	Weekly		X	X	X	X		X
Inside	Weekly		X	X	X	X		X
Glass	Daily		X	X	X			X

Source: Naumann *et al.* (1964).

SPOILAGE FLORA

Mossel and Ingram (1955) listed *Pseudomonas, Flavobacterium, Micrococcus, Cladosporium* and *Thamnidium* as the dominant spoilage association under normal conditions of storage. *Achromobacter* which was listed by Mossel and Ingram in this group is not considered a valid genus according to the latest *Bergey's Manual of Determinative Bacteriology.* These Gram-negative rods are now classified as *Alcaligenes* sp. Lechowich (1971) found lactobacilli and microbacteria associated with spoilage of fresh meat. He noted that *Pseudomonas* and *Flavobacterium* sp. cause slime, greenish discoloration, fluorescent pigments and white-to-colored spots, whereas *Lactobacillus, Micrococcus* and *Microbacterium* produce sliminess or stickiness, souring or putrefaction.

SLAUGHTERING BEEF

Premortem. In addition to the effects of excitement on glycogen utilization with a resulting rise in pH of the meat, cattle should not be excited. Such cattle do not bleed well; this lowers the keeping quality of the meat. Also, the cattle should not be fed 24–48 hr prior to slaughter, since fasted animals bleed out more thoroughly (Romans and Ziegler 1974).

Stunning. Stunning of the animal may be achieved by compression guns. By stunning, there is less struggle during bleeding and less use of glycogen.

Sticking. Prior to sticking, the animal is hung on an overhead rail. The carotid arteries and jugular vein are cut, allowing bleeding to occur.

The blood is collected to prevent its entrance into the sewage system.

Skinning. The animal can be skinned on the rail or a rack. See Romans and Ziegler (1974) for details.

Eviscerating. The body cavity is opened, and the viscera are removed and inspected for diseased parts by a veterinarian. Diseased animals are eliminated in premortem examinations.

Handling Practices During Skinning and Eviscerating. The microflora on the hide of an animal may range from 10^5 to 10^6 aerobes per square centimeter of unwashed surface. If the hide had manure present, the counts would be very high. Extreme caution must be taken to prevent the transfer of microorganisms to the carcass. Cattle and sheep are skinned before they are eviscerated. Care must be exercised to prevent cutting of an intestine and contaminating the meat with billions of bacteria. Also, bacteria may be transferred from soil on hooves and even from the workers themselves.

Washing the Carcass. The carcass is split, usually with the aid of an

TABLE 7.4.   COOLING RATE OF BEEF CARCASSES

Time in Hours	Round Avg Temp (°C)	(°F)	Chuck Avg Temp (°C)	(°F)
0	39.4	103	37.8	100
2	38.3	101	31.1	88
4	33.9	93	23.9	75
6	29.4	85	20.6	69
22.5	11.1	52	3.3	38
24.5	9.4	49	2.8	37
26.5	7.2	45	1.7	35
29.5	5.5	42	1.7	35
32.5	3.3	38	1.7	35

Source: Jensen (1942).

electric saw, and the sides washed with high pressure, potable water. Chlorinated water could be used in the wash to reduce the number of surface microorganisms. After washing, a wet muslin sheet (called shroud) is wrapped tightly around each half of the carcass. Shrouds prevent some loss of moisture and smooth out the fat on the carcass. Shrouds should be clean and handled in a sanitary manner to prevent heavy contamination of the meat.

Cooling of Beef Carcasses. Jensen (1942) conducted tests to determine the rate of cooling on 48 carcasses of beef held at 0°–1.6°C (32°–35°F). The temperatures taken from the round and chuck areas are given in Table 7.4. Fresh meats must be chilled as rapidly as possible to prevent mesophilic microorganisms from growing during this cooling period. If the population can be kept in the lag phase of their growth curve, the meat will remain sound.

Placing of "hot" or "just-killed carcasses" in with cooled carcasses will cause moisture deposit on the cooled carcasses. As shown in Table 7.5, wet surfaces are more favorable for microbial growth.

Aging Beef. During tenderizing of meat in controlled humidities and

TABLE 7.5.   EFFECT OF WETTING ON STORAGE OF LAMB CUTS

	Aerobic Bacteria per Gram-Nutrient Agar with Initial Count 25,000 to 30,000 per Gram			
	Held at 2.2°–3.3°C (36°–38°F)		Held at 7.2°–10.0°C (45°–50°F)	
Time in hr	Wet Surface	Dry Surface	Wet Surface	Dry Surface
24	400,000	40,000	1,000,000	200,000
48	660,000	45,000	6,000,000	. . . . .
72	760,000	42,000	putrid	4,000,000 st. off-odor
96	off-odor 2,000,000	sweet 70,000	putrid	3,000,000 st. off-odor

Source: Jensen (1942).

temperature, various molds may develop on the carcass. These "whiskers" may be *Thamnidium, Mucor* and *Rhizopus.* Ultraviolet light may aid in control of molds.

Spoilage (Odor and Slime). Between 1.2 and $100 \times 10^6/cm^2$ microorganisms are needed to detect odor on beef, whereas between 3 and $300 \times 10^6/cm^2$ microorganisms are present when slime is evident (Frazier 1967).

White, green, yellow, greenish-blue to brown-black spots and purple discoloration may occur on beef due to the growth of microorganisms. *Pseudomonas* predominates on beef held at 10°C (50°F) or lower. Frazier (1967) observed that micrococci occur at temperatures above 15°C (59°F).

Yellow discoloration may be due to *Flavobacterium.* Greenish-blue to brown-black spots are caused by *Cladosporium.* Greenish discoloration may be caused by (1) alpha streptococci, (2) hydrogen sulfide-forming bacteria and (3) excessive contamination with certain species of *Pseudomonas.*

SLAUGHTERING HOGS

Antemortem Conditions of Animals. The same principles apply to hogs that pertain to beef cattle in relation to their health, state of stress and possible contamination.

Immobilization. Mechanical, chemical and/or electrical devices may be used. As with beef, the immobilization process should minimize excitement and discomfort.

Sticking. The hogs are shackled and hoisted to a rail, stuck and bled. The total number of bacteria on the 2 sq in. of neck skin where jugular vein and carotid arteries are cut may vary widely as shown in Table 7.6.

The microflora of the hog, as with the microflora of cattle, is mixed. The

TABLE 7.6. BACTERIA ON TWO SQUARE INCHES ON NECK SKIN WHERE JUGULAR VEIN IS STUCK

Hog	Aerobes	Total Numbers (Unwashed Hogs) Anaerobes
1	500,000,000	200,000,000
2	100,000	1,000,000
3	1,500,000,000	2,000,000,000
4	500,000	10,000
5	240,000,000	190,000,000

Source: Jensen (1942).

habits of the hog are such that the skin of the hog can become grossly contaminated. During the sticking operation, the sticking knife carries microorganisms into the blood. During the 3–4 min while the hog is bleeding to death, bacteria penetrate the hog's carcass. This period is called the agonal invasion period.

Scalding. Water for scalding should be 65.5°C ± 5° (150°F ± 5°). Alkalies are added to the water to loosen the hair and scurf. Hair is removed with a bell-type scraper (Romans and Ziegler 1974).

Eviscerating. The body cavity is opened; the viscera are removed and inspected. The carcass is split and washed thoroughly with potable water.

Prechill. Hot carcasses are placed in a prechill so that moisture will not be transferred to cooler carcasses. The temperature in this room may be a few degrees below freezing to counteract the hot carcasses.

Chill. After the preliminary chilling, the cooled carcasses are transferred to a chill room where the temperature may be held around 1.1°C (34°F) (Romans and Ziegler 1974).

PROCESSED COMMINUTED PRODUCTS

Other Than Sausages

Hamburger. The microbial quality of hamburger is influenced by the numbers and kinds found on the meat and on the equipment used in this preparation. Ayers (1973) stated that the initial microbial loads of freshly ground beef, even when carefully handled, are in the range of 100,000 per gram. As meat is cut and passed through the grinder, the load becomes redistributed. Of course, a grinder which is not cleaned thoroughly may contain bacteria to seed meat with a low count. Counts of 60–100 million per cm^2 may be detected by the time slime is noted. *Pseudomonas* is the dominant bacterium at this time. Off-odor occurs before slime production but by only a few hours.

Canada's Health Protection Branch is proposing mandatory standards for the microbiological quality of ground beef. The suggested standard is as follows (Anon. 1975):

(1) 10,000,000/g total aerobic count for unfrozen meat.
(2) 1,000,000/g total aerobic count for frozen meat.
(3) *Escherichia coli* count of 100/g.
(4) Coagulase-positive staphylococci count of 100/g.
(5) *Salmonella* 0 count in 25 g.

Sausages (General)

According to Romans and Ziegler (1974), certain equipment is essential in large-scale production of sausages. All machinery should be kept scrupulously clean to prevent contamination of the product with spoilage bacteria. Grinders, silent cutter or chopper, stuffer, linker, peeler, smoker, and cooler are used. A colloid mill (a type of silent cutter) may be employed to prepare an emulsion of the ingredients.

Lechowich (1971) listed the following microorganisms as spoilage agents on "cured" sausages:

> Surface slime—*Micrococcus* and yeasts
> Gas production in vacuum-packed frankfurters—*Lactobacillus*
> Fading of cured meat at outer surfaces—*Leuconostoc* and *Micrococcus*
> Greenish discoloration—*Lactobacillus*

He found that yeasts and molds caused slime and discoloration on fermented sausages.

Cured Sausages (Frankfurters, Bologna). These products are usually a mixture of pork, beef, sugar, salt, sodium nitrite and/or sodium nitrate and spices. The meat is ground finely in a slurry. The raw materials need to be of high quality to ensure a good product as the bacteria are incorporated throughout the mixture. The pasty mixture may be pumped and extruded onto rubberoid conveying belts and carried to stuffers where it is inserted into casings, and then cooked, smoked, cooled and packaged.

Water activity in meat products influences the growth of microorganisms. This is illustrated in Table 7.7.

Vacuum packaging of frankfurters stimulates the growth of facultatively anaerobic yeasts and heterofermentative lactic acid bacteria. Yeast produces surface slime, whereas lactic acid bacteria can produce swells due to formation of CO_2 (Lechowich 1971).

Greening is a defect in sausages caused by microorganisms. The discoloration which occurs as a green core or as a green surface may be caused by heterofermentative species of *Lactobacillus, Leuconostoc,* catalase-negative bacteria and hydrogen sulfide producers (Frazier 1967). Frazier stated that greening is favored by a slightly acid pH and by the presence of small amounts of oxygen.

If a cut surface is green, this indicates contamination with and growth of salt-tolerant, peroxide-forming psychrophilic bacteria (probably lactics). Surface slime may accompany the greening. Slime formation may be spread from sausage to sausage.

Summer Sausages. Summer or farmer-style sausage is made from pork,

TABLE 7.7. a_w OF MEAT PRODUCTS AND a_w LIMITING THE GERMINATION OR GROWTH OF MICROORGANISMS

Product	a_w	Microorganism
Distilled water	1.00	
Fresh meat[1]	0.99	Bacillus mycoides[2]
	0.98	
	0.97	Pseudomonas[3]
	0.96	E. coli[3]
	0.95	Bacillus subtilis[3]
	0.94	
Frankfurter[1]	0.93	Mucor, Rhizopus, Botrytis[3]
Sausage[1]	0.92	
Sliced pressed ham[1]	0.91	Micrococcus roseus[2]
Sliced pressed bacon[1]	0.90	Yeast from condensed milk[3]
Cured pork[1]	0.89	
	0.88	
Smoked pork[1]	0.87	
	0.86	Staphylococcus aureus[3]
	0.85	Aspergillus spp.[3]
	0.84	Alternaria citri[2]
	0.83	
	0.82	
	0.81	
Fermented sausage with high salt content[1]	0.80	Penicillium sp.[2]

[1] According to Hansen and Riemann (1962).
[2] According to Scott (1957).
[3] According to Frazier (1967).

beef, salt, sugar, pepper, mustard seed, and curing agents such as nitrite and nitrate. The meat is ground and mixed with the other ingredients and refrigerated for several days. The mixture is then stuffed into hog casings and smoked; the temperature reaches 61.1°C (142°F). The sausages are stored in a dry, cool place (4.4°–10°C; 40°–50°F) for several weeks before they are used (Romans and Ziegler 1974). During this time naturally-occurring lactic acid bacteria in the meat develop and produce the tang of this typical fermented sausage.

Pedicoccus cerevisiae may be added to ensure a desired flavor.

Dipping or spraying the casing with 2% sorbic acid solution will prevent mold growth.

PROCESSED NONCOMMINUTED PRODUCTS

Country-Cured Hams (Dry Cure). A mixture of 7 lb of salt, 3 lb sugar (white or brown) and 3 oz of sodium nitrate is a basic cure for country-cured hams (Alexander and Stringer 1972). Salt is the primary ingredient used in meat curing. According to Rust and Olson (1973), sodium chloride serves as a preservative for "country style" cured meats such as dry and semidry sausages and hams. To function as a preservative, a salt concentration of about 17% is needed. In most modern- or quick-cured hams, a salt concentration of 2–3% is used.

Sodium nitrite is bacteriostatic. This compound is an effective inhibitor against clostridia, especially *Clostridium botulinum.* Since country curing is a slow process, sodium nitrate may be used since the nitrate is reduced to nitrite. Therefore, the role of nitrates is in fermented sausages, dry-cured hams and bacon.

In a long process, like the dry cure, sugar provides an energy source for the bacteria that are reducing the nitrate to nitrite. Some of the sugar ends up as lactic acid which is produced by the lactic acid bacteria.

Rust and Olson (1973) state that 40 days are a minimum of time for the center of dry-cured hams to approach 1% of the curing chemicals in the meat. Penetration by the curing mixture, a function of time and temperature, is more rapid at higher temperatures, with an upper limit of about 4.4°C (40°F) because of spoilage bacteria.

The mixture is applied to the surface of the ham at one, two or three 5-day intervals (Rust and Olson 1973). A dry-cured product will use upwards to 3% of the curing mixture based on the original weight of the ham.

Country-cured hams being produced by a processor near Kansas City are aged in rooms with a relative humidity of 75% for about 60 days. Mold control, a possible problem during this aging process, can be achieved by using sorbic acid on the paper and cloth bags used to hang the hams. Proper air circulation and proper sanitation will lower the mold problem. The use of sodium hypochlorite to scrub down ceiling, walls, floors and the wooden holding racks can also help.

Sutic *et al.* (1972) studied 562 molds isolated from country-cured hams; 403 isolates were from the genus *Penicillium;* 121 were from *Aspergillus*; and 36 were *Cladosporium, Alternaria* and other genera. Two strains of *Aspergillus flavus* produced aflatoxin. In a study in 1973, Escher *et al.* found that two strains of *A. ochraceus* produced ochratoxin A and B. After 21 days, $2/3$ of the toxin was found penetrating the meat to a distance of 0.5 cm.

In addition to ochratoxin, Ayres (1975) found that six strains of *Penicillium viridicatum* produced the mycotoxin, citrinin. Strains of *Aspergillus versicolor,* isolated from cured hams, produced sterigmatocystin, another mycotoxin.

Ayres (1975) tested cultures of molds isolated from aged cured meats for their toxicity to chick embryos. These results are given in Table 7.8. Considering the long exposure time to molds for country-cured hams, every effort should be made to control these fungi. If they do not grow on the surface, there will be no toxin production within the meat.

TABLE 7.8.   TOXICITY OF MOLDS ISOLATED FROM AGED, CURED MEATS

Species	Toxicity to Chick Embryos
Aspergillus ruber	2/22
Aspergillus repens	5/28
Aspergillus sydowi	2/12[1]
Aspergillus restrictus	1/12
Aspergillus amstelodami	2/7
Aspergillus chevalieri	1/2
Aspergillus fumigatus	1/1
Penicillium expansum	2/15
Penicillium notatum	1/3
Penicillium brevi-compactum	1/2
Penicillium sp.	1/8

Source: Ayres (1975).
[1] A thousand-fold dilution of one strain inoculated into the egg air sac produced 0% hatch.

Quick-Cured Hams. The major difference in "quick-cured hams" and "country-cured hams" is the speed at which the cure reaches all parts of the meat and the salt concentration used which is high in country-cured hams and low in quick-cured hams. In country-cured hams, the cure must diffuse into the ham, whereas the cure is pumped into the ham in quick curing. These hams need refrigeration, whereas country-cured do not; however, mycotoxins develop in country-cured hams but not in quick-cured hams. Food poisoning is more of a threat with the quick cure than with the country cure because of the growth of food poisoning bacteria in improperly handled quick-cure hams.

Bacon. In the manufacture of bacons, the cure materials are injected into the belly portion of the slab at multiple places to speed the diffusion of the brine throughout the meat.

Smoking and Heating. Processed meats are placed in a smokehouse to cook at temperatures up to 65°–75°C (149°–167°F). This is enough heat to kill most nonsporeforming microorganisms as well as trichinae (Forrest *et al.* 1975). The meat is essentially pasteurized by this process.

Smoke imparts flavor and inhibits the growth of microorganisms. More than 200 compounds have been identified in smoke. Some examples are aldehydes, ketones, alcohols, phenols, organic acids, cresols and acyclic hydrocarbons. Some of these compounds have bacteriostatic or bactericidal properties, but it is believed that formaldehyde accounts for the preservative action (Forrest *et al.* 1975).

Cured Beef Tongues. Small pieces of meat may be cured by brine soaking. Larger pieces of meat might spoil before the brine penetrates throughout. The strength of the brine varies with the desired saltiness. Tongues can also be pumped with a pickling liquid. A sweet pickle might consist of 12.5% salt, 1% sugar, sodium nitrate and nitrite at 0.08%.

Dried Beef or Beef Hams. Boneless cuts derived from the hind leg of the beef carcass are used as the starting material. The pickle is pumped into the meat which is then held in a cover pickle. After curing, the meat is smoked, sliced and packaged. Counts on the fresh meat at the start of the process should be under 500,000 per gram. Spoilage of dried beef in jars may be due to denitrifying aerobic bacteria or *Pseudomonas fluorescens* may produce gas from oxides of nitrogen. Jensen (1942) found that *Bacillus* sp. can produce CO_2 in jars of dried beef and may also cause sour rounds of beef.

MICROBIOLOGICAL ANALYSES

According to Ayres (1973), the most important microbiological tests are (1) total numbers, (2) coliforms, and (3) coagulase-positive staphylococci. He thinks that it would be interesting to know if the sample contained salmonellae, shigellae, anaerobes, molds and yeasts, but such additional tests are too costly.

Total counts give useful information on whether or not the meat contains unusually high counts for a quality product. The total count can give information on handling practices and sanitation.

Coliforms also give data on the sanitary conditions during production. Usually this type of examination consists of incubation of the food in lactose broth and then streaking bacteria from positive tubes (as indicated by gas production) on eosin methylene blue or endo medium plates. Positive plates are indicative that coliform bacteria are present. Positive coliforms indicate (1) that the product may have been exposed to fecal contamination, (2) that unfit raw material may have been used, or (3) that the product has received unsanitary handling.

Ayres (1973) gives the following pattern for testing for coagulase-positive staphylococci:

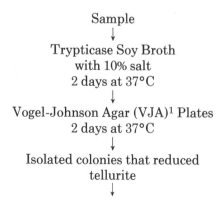

Sample
↓
Trypticase Soy Broth
with 10% salt
2 days at 37°C
↓
Vogel-Johnson Agar (VJA)[1] Plates
2 days at 37°C
↓
Isolated colonies that reduced
tellurite
↓

[1] See Chap. 17 for description of VJA.

Agar slants
↓
Brain heart infusion (BHI broth)
18–24 hr at 37°C
↓
Add 1/2 ml of reconstituted
coagulated plasma with
EDTA
↓
Clot formation within
4-hr period

GLOSSARY

Acyclic hydrocarbons—These hydrocarbons have an open chain molecular structure rather than a ring-shaped structure.

Aflatoxin—The toxin is produced by several fungi, including strains of *Aspergillus flavus*. There are several kinds of aflatoxins.

Agonal invasion—Refers to the time after cutting the artery in the neck of an animal until the animal dies. Bacteria may circulate throughout the animal's body if they were on the knife used in sticking.

Antemortem or premortem—Before death; postmortem, after death.

Coagulase positive—Coagulase positive means that the bacterium produces the enzyme, coagulase, that coagulates blood serum.

Coliforms—Bacteria from both plant and animal sources pertain to or resemble the colon bacillus. *Escherichia coli* and *Enterobacter aerogenes* are coliform bacteria.

Comminuted products—To reduce to a powder, to pulverize. In meat products, the meat is ground to make a smooth slurry as in the manufacture of sausages.

Country-cured hams—This process uses a high salt concentration and may take as long as 60 days to complete. These hams do not need refrigeration as do the quick-cure hams.

Cresols—Refers to any of three isomeric phenols used as a disinfectant. As used in this chapter, however, this compound is part of the smoke. Phenols are also used as disinfectants.

Curing—A process for preserving meat and fish by pumping a cure mixture (salt, sodium nitrate, sodium nitrite, sugar) throughout the foodstuff. The product is then smoked and aged.

Eviscerate—To remove the entrails or to disembowel.

Glycogen—This is the chief means for storage of carbohydrate by animals. It is also called animal starch. Glycogen is stored in the muscle and liver.

Mycotoxins—Toxins produced by fungi.

Ochratoxin—A special kind of mycotoxin which is produced by *Aspergillus ochraceus*.

Pickle—The mixture of salts and sugar used in curing meats.

Potable—Water that is drinkable or water free of all harmful chemicals and microorganisms.

Prechill—The immediate cooling of hot carcasses before storage in chill rooms.

Premortem—See antemortem.

Proteolytic—Having the power to break down proteins into simpler or more soluble substances.

Quick-curing—The pickle contains less salt so that the mean product must be refrigerated.

Saltpeter—Potassium nitrate.

Sanitizers—Compounds used to reduce the microflora to low levels.

Scalding—To pass the carcass of an animal through warm water to help remove the hair.

Shroud—A cloth used to wrap one half of a beef carcass.

Sticking—Cutting of the artery in the bleeding process.

Summer sausage—A fermented (lactics) meat product.

Sweet pickle—A curing mixture containing sugar.

REVIEW QUESTIONS

1. Why should animals be rested before slaughter?
2. What is the pH of meat after slaughter?
3. In what form is energy stored in muscle?
4. List possible sources of contamination during slaughter and handling.
5. How good is meat as a substrate for microbial growth?
6. How often should equipment used in meat handling be cleaned?
7. List microorganisms associated with spoilage of fresh meat.
8. Describe the slaughter of beef and hogs and explain how the process influences the microbiological quality of the meat.
9. Why should hot carcasses not be put in a cooler with cold carcasses?
10. List microorganisms which grow and produce colored spots on meats.

11. What are the counts which correspond to slime? To off-odors?
12. What is the suggested standard for hamburger?
13. What influence does vacuum packing have on the microflora of meat?
14. Give some examples of cured sausages.
15. What causes greening in sausages?
16. What is summer sausage?
17. How can mold be prevented from growing on summer sausage?
18. What is the function of salt, sugar, sodium nitrate and sodium nitrite in curing country hams?
19. What effect does sodium nitrite have on spoilage bacteria?
20. What are mycotoxins? Are there dangers of mycotoxins contaminating country cured hams?
21. How can mold be controlled during curing of country hams?
22. Name four mycotoxins produced by molds (growing on country cured hams).
23. What two genera of molds dominate on country-cured hams?
24. What is one way of testing for mycotoxin production?
25. What is a quick-cured ham?
26. Should quick-cured hams be refrigerated?
27. What do smoking and heating cured meats do to the microflora?
28. What are three ways curing materials may be applied to meat?
29. What are the three microbiological tests Ayres recommends for meat products?
30. Describe how to test for coliforms and coagulase-positive staphylococci in meat products.

REFERENCES

ALEXANDER, M. A. and STRINGER, W. C. 1972. Country Curing Hams. Univ. Missouri-Columbia Guide.

ANON. 1975. New Notes Washington. J. Food Technol. *29*, 112.

AYRES, J. C. 1973. Microbiology of cured meats. Proceedings 15th Annual Meat Science Institute, Univ. Georgia.

AYRES, J. C. 1975. Toxigenic fungi in food products and feeds. Research grants supported by the Food and Drug Administration, July 1, 1969–June 30, 1975. Available from Food and Drug Admin., Office of Science, U.S. Dept. of Health, Education and Welfare, Public Health Service, Rockville, Maryland.

ESCHER, F. E., KOEHLER, P. E. and AYRES, J. C. 1973. Production of ochratoxins A and B on country-cured hams. Appl. Microbiol. *26*, 27–30.

FORREST, J. C. *et al.* 1975. Principles of Meat Science. W. H. Freeman and Co., San Francisco.

FRAZIER, W. C. 1967. Food Microbiology. McGraw-Hill Book Co., New York.

HANSEN, N. H. and RIEMANN, H.   1962.   Microbiological quality of packaged meat and meat products.   Fleischwirtschaft *14*, 861. (German)

JENSEN, L. B.   1942.   Microbiology of Meats.   Garrard Press, Champaign, Illinois.

LECHOWICH, R. V.   1971.   Microbiology of meat.   *In* The Science of Meat and Meat Products.   J. F. PRICE and B. S. SCHWEIGERT (Editors).   W. H. Freeman and Co., San Francisco.

MOSSEL, D. A. A. and INGRAM, M.   1955.   The physiology of the microbial spoilage of food.   J. Appl. Bacteriol. *18*, 231–268.

NAUMANN, H. D., STRINGER, W. C. and GOULD, P. F.   1964.   Handling prepackaged meat.   Manual *64*, Ext. Div. Univ. Missouri.

ROMANS, J. R. and ZIEGLER, P. T.   1974.   The Meat We Eat.   Interstate Printers and Publisher, Danville, Illinois.

RUST, R. E. and OLSON, D. G.   1973.   Meat curing principles and modern practice.   Koch Supplies, Kansas City.

SCOTT, W. J.   1957.   Water relations of food spoilage microorganisms.   Adv. Food Res. *7*, 83, Academic Press, New York.

SUTIC, M., AYRES, J. C. and KOEHLER, P. E.   1972.   Identification and aflatoxin production of molds isolated from country cured hams.   Appl. Microbiol. *23*, 656–658.

WATT, B. K. and MERRILL, A. L.   1963.   Composition of foods.   USDA Agric. Handbook *8*.

WEISER, H. H., MOUNTNEY, G. J., and GOULD, W. A.   1971.   Practical Food Microbiology and Technology, 2nd Edition.   AVI Publishing Co., Westport, Conn.

8

Microbiology of Poultry Meat

POULTRY MEAT AS A SUBSTRATE FOR MICROORGANISMS

The chemical composition of both light and dark raw chicken meat is given in Table 8.1.

In review of factors influencing the growth of spoilage microorganisms, chicken meat is a good source of protein; therefore, the spoilage microflora are proteolytic. They obtain both carbon and nitrogen from the protein present in the substrate. Although the flora may also degrade or synthesize fats, microorganisms do not need these substances for growth.

Data in Table 8.1 also show that poultry meat is a good source of vitamins and minerals (in the ash).

The same basic principle of rested vs excited birds applies here, as it did in beef and pork. Glycogen is stored in the muscle. After slaughter, this will be converted into lactic acid and will drop the pH. If all or most of the glycogen is used up in excited activity, the pH will remain higher after death. Weiser *et al.* (1971) listed the pH of chicken as being 6.2–6.4. This is a very good pH to support extensive microbial growth.

Since the substrate is excellent for growth, temperature and sanitation must be used to control the growth of spoilage bacteria during handling and storage of raw poultry.

SPOILAGE FLORA

According to Mossel and Ingram (1955), *Pseudomonas, Flavobacterium* and *Micrococcus* are the genera dominating when spoilage occurs during standard conditions of storage.

The numbers of bacteria vary per bird, but studies show that counts can be as low as 1500 per sq cm of the skin surface of live birds and 35,000 per

TABLE 8.1.   CHEMICAL COMPOSITION OF CHICKEN[1]

Type of Meat	Water (%)	Protein (%)	Fat (%)	Ash (%)	Thiamin (mg/100g)	Niacin (mg/100g)	Riboflavin (mg/100g)
Light, without skin, raw	73.7	23.4	1.9	1.0	0.05	10.7	0.09
Dark, without skin, raw	73.7	20.6	4.7	1.0	0.08	5.2	0.20

[1] Chicken meat contains no carbohydrate.
Source: Watt and Merrill (1963).

sq cm on birds immediately after processing. According to Dawson and Stadelman (1960), a population of 10^7 per sq cm indicates spoilage. Chickens with counts of 10^8 have off-odors and slime. After storage of fresh poultry meat at refrigerated temperatures, *Pseudomonas* will account for 90–95% of the microbial flora.

COMMERCIAL POULTRY PROCESSING

The monograph, *The Bacteriology of Commercial Poultry Processing* by Gunderson *et al.* (1954) presented a typical, approved, poultry kill and eviscerating line. In addition, there are two papers in *Applied Microbiology* which show floor diagrams of poultry processing plants which you may also wish to study (Kotula and Kinner 1964; Morris and Wells 1970).

Slaughtering and Eviscerating Poultry

The following steps are recommended in preparing poultry for market.

Premortem Handling. Care should be used to prevent bruises, abrasions and broken limbs. Poultry should be held in clean cages at the plant. The room should be cool and well ventilated. The birds should have water but no food 8 to 10 hr before slaughter (Romans and Ziegler 1974). This helps in maintaining sanitary conditions on equipment during processing. Poultry full of feed will not bleed well.

Only healthy birds should be processed.

Killing and Bleeding. Cutting the throat inside the bird's mouth is the approved and most widely used method today (Romans and Ziegler 1974). Bleeding cones or an overhead conveying system may be used to ensure adequate bleeding and a high-quality product. The blood is caught and removed separately so it does not overload the sewage system.

Scalding. Scalding to facilitate the removal of the feathers may be accomplished by immersion in water at 53°–59°C (128°–138°F) or by spraying the birds with water at 59°–61°C (138°–142°F). Both operations take 30 sec. A reasonable turnover in water is needed during the scalding process to eliminate gross contamination of the scald water when processing hundreds of birds.

Singeing. Singeing is a process used to remove fine hair and residual feathers. Of course, this procedure reduces somewhat the numbers of surface bacteria.

Washing Prior to Evisceration. Washing of the carcass removes dirt, debris, and the bacterial load. Chlorinated water should be used.

Singeing. The birds are singed again which aids in drying skin surfaces, destroying some bacteria and removing some burnable debris.

Removal of Pin Feathers. Workers check the carcass for and remove any remaining pin feathers.

Evisceration. (a) The preen gland is removed. Cuts are made around the neck, and skin is stripped from the neck. The position of the bird is changed so it hangs from the head, not the feet.

(b) An incision is made around the thighs and vent attempting not to cut the intestine. (A cut in the intestine would contaminate the carcass with millions of bacteria.) From the standpoint of sanitary control of product, this is an important step.

(c) The viscera are removed from the bird. Workers handle both the carcass and the viscera at this point. Obviously, there are large numbers of bacteria on the hands of the workers due to fecal soilage from the birds.

(d) The carcass and viscera are inspected by a veterinarian (either USDA or State-appointed).

(e) Giblets are removed to a side table where they are washed, cleaned, and packaged for frozen storage.

(f) Kidneys, ovaries, and remnants of lung tissue are removed from the body cavity.

(g) Head and neck are removed.

(h) Washing of the body cavity is the next step.

(i) Final inspection.

(j) Final wash.

Types and Numbers of Bacteria on Poultry on Evisceration Line (Gunderson *et al.* 1954)

Counts of 26,000 per sq cm were found on birds arriving at the eviscerating line, whereas counts averaged 3500 per sq cm on completing evisceration.

Coliforms and beta-hemolytic bacteria counts averaged 3000 and 1000 per sq cm, respectively, at the start of evisceration and averaged 300 and 80 after evisceration.

Efficient washing of the carcasses reduced bacterial counts; handling procedures caused an increase in bacterial population.

Of 2184 cultures isolated, 261 were salmonellae. The following salmonellae which have been incriminated as agents of human disease were found: *S. montevideo, S. typhimurium, S. panama, S. sentenberg, S. gallinarum, S. cholerae-suis* var. *kuzendorf.*

THE *SALMONELLA* PROBLEM (FOOD INFECTION)

In a study of the survey of market poultry for *Salmonella* infection, Sadler *et al.* (1960) found that these Gram-negative rods were usually associated with disease processes in fryer chickens but seldom in turkeys. Both infected and carrier birds were found. More infected turkeys (43% of the sampling days) were found than chickens (26% for chickens and 17% for hens). *S. typhimurium*, the most common of salmonellae causing food infection, was found in intestines of both chickens and turkeys.

Salmonella was isolated from broiler-fryer type chickens purchased in retail stores in Lafayette, Indiana in 1963 (73 out of 264 chickens were positive). Woodburn (1964) found in this study 13 different serotypes, the most common being *S. infantis, S. reading,* and *S. blockley.*

More recently, Morris and Wells (1970) reported 202 (14.2%) of 1427 samples from a poultry-processing plant over a 2-year period were positive for *Salmonella.* They stated that recontamination occurred during eviscerating and chilling. They found that 48/203 of the carcasses were positive samples initially. After evisceration, salmonellae were found in 38/212 carcasses and in 26/108 edible viscera (giblets). In testing, plant environment (tubs, conveyers, knives, saws, and gutter water) they found 28/100 sites were positive for salmonellae.

Not only is there a potential danger of poultry products having *Salmonella,* but also there is evidence of these causing food infections. In 1973, there were 33 incidents involving salmonellae in foods. The six outbreaks which involved chicken or turkey are shown in Table 8.2.

FOOD POISONING PROBLEMS

In addition to the *Salmonella* problem, poultry acts as a substrate for other foodborne disease. Examples listed in Table 8.3 indicate that food poisoning can occur at home, at a church, or a lodge, and even in the hospital.

TABLE 8.2.   LISTING OF *SALMONELLA* OUTBREAKS IN 1973

Etiology	State	No. of Cases	Food Vehicle	Location Where Food Was Mishandled
S. thompson	California	33	Chicken molé Potato salad	Home
S. chester	California	66	Turkey	Church
S. enteritidis	Massachusetts	24	Chicken salad	Party room
S. manhattan	Oregon	60	Turkey	Fraternity house
S. agona	Pennsylvania	142	Chicken	Restaurant
S. reading	Virginia	470	Turkey salad	School

Source: Annual Summary of 1973 published by Center for Disease Control (1974).

SANITATION

Important to all food manufacturing plants are an adequate building and equipment which is easy to keep clean. Furthermore, the plant and its equipment should be insect- and rodent-proof. In their study, Gunderson *et al.* (1954) found that in-plant chlorination was essential in maintaining a clean and odor-free plant. Specifically, they found the following:

(1)   Levels of 10 and 20 ppm residual chlorine were satisfactory for in-plant chlorination.
(2)   Use of in-plant chlorinated water sprayed on birds and equipment reduced the microflora on poultry.
(3)   Clean-up of the plant was easier, requiring less labor and time.
(4)   Levels of chlorine used did not cause any off-odor in the poultry meats.
(5)   Chlorine was effective and economical.

TABLE 8.3.   LISTING OF FOODBORNE DISEASE OUTBREAKS

Etiology	State	No. of Cases	Vehicle	Location Where Food Was Mishandled and Eaten
Clostridium perfringens	California	51	Chicken	Convalescent hospital
C. perfringens	Tennessee	800	Turkey	Cafeteria
C. perfringens	Utah	11	Turkey	Home
C. perfringens	Washington	33	Turkey	Lodge
Staphylococcus aureus	California	6	Turkey	Home
S. aureus	Oklahoma	80	Turkey	Church
S. aureus	Washington	3	Barbecued chicken	Home

Source: Annual Summary of 1973 published by Center for Disease Control (1974).

POULTRY PRODUCTS

Meat Pies

Frozen meat pies have been on the market for many years. Deboned poultry products are used (chicken and turkey). The deboning process and handling can add considerable numbers of bacteria to the meat in the product if not handled properly.

Gunderson *et al.* (1954) listed results of bacteria counts in frozen chicken pot pies following cooking (Table 8.4). The data show variation between samples. Kereluk and Gunderson (1960) used heavy suspensions of *S. faecalis* to contaminate frozen meat pies artificially. In this study, the pies were baked at 218°C (425°F). After 30 min, nonsporeforming bacteria were reduced to 0–3.6% survivors, whereas after a 20-min baking period, nonsporeforming bacteria had a survival rate of 14 to 100%. (Nonsporeformers included *Staphylococcus aureus*, *S. faecalis* and *Escherichia coli*.) Although meat pies are a potential source of food poisoning, few outbreaks have been traced to them. However, an instance of botulism caused by a commercially processed frozen chicken pie is listed in Chap. 17 (Table 17.10).

TABLE 8.4.   BACTERIAL PLATE COUNTS OF FROZEN PACKAGED CHICKEN POT PIES (FOLLOWING COOKING)

Sample No.	Total Viable Bacteria per Gram	Total Coliform Bacteria per Gram	Total β-Hemolytic Bacteria per Gram
1735	62,000	150	7,500
1736	405,000	50	TNTC[1]
1737	12,000	180	900
1738	5,900	1,100	850
1739	13,400	290	580
1740	145,100	40	TNTC[1]
1741	22,100	180	6,000
1742	158,200	3	570
1743	2,700	35	1,200
1744	5,700	50	4,500
1745	25,700	80	430
1746	10,300	40	270
1747	30,500	60	450
1748	49,500	35	380
1749	15,300	70	TNTC[1]

Source: Gunderson *et al.* (1954).
[1] Too Numerous To Count.

Turkey Rolls

Turkey rolls are a precooked product that are popular today. Mercuri *et al.* (1970) studied the bacteriological condition of commercially cooked Eastern-type rolls during processing and storage. They evaluated the products at various times under refrigeration at 5°C to simulate handling and storage practices of distributors and institutional users. After 2 weeks at 5°C aerobes on the surface of rolls, in slices and in whole rolls numbered 1 to 10^6 per cm^2 or per gram. Coliform bacteria counts increased from 10^5 to 10^6 per gram and enterococcus counts rose from $<10^2$ to more than 10^6 per gram. Also, 8 of 28 rolls (after handling and packaging) contained coagulase-positive staphylococci. They occasionally found salmonellae.

Mercuri *et al.* (1970) concluded that strict sanitation principles and practices were necessary during handling and packaging of the rolls to minimize contamination of the cooked product with foodborne pathogens. Since heat-treating meats reduces the natural, competing microflora, pathogens are able to reproduce more readily and thus cause food poisoning.

GLOSSARY

Bleeding cones—Metal cones (stainless steel) into which the bird is placed during bleeding.

Broiler—A tender young chicken.

Giblets—The heart, liver and gizzard of a fowl.

Pin feathers—Small feathers occurring on fowl.

Scalding—Treating the fowl with warm water to facilitate the removal of feathers.

Turkey roll—Deboned turkey meat made into a roll.

Vent—The anal excretory opening in fowl and other birds.

Viscera—The intestinal organs of the body.

REVIEW QUESTIONS

1. Does poultry meat make a good substrate for the growth of spoilage flora? Why?
2. From the factors needed for microbial growth, compare poultry meat as a substrate with other meats.
3. List the spoilage flora that dominate during normal conditions of storage.
4. Describe slaughtering and eviscerating poultry. What factors during slaughtering and eviscerating add to and/or reduce the bacteria on the product?

5. Why do you think salmonellae contaminate poultry carcasses?
6. Of what significance are salmonellae on poultry meats?
7. Name two food poisoning bacteria associated with poultry or poultry products.
8. Which sanitary practices should be followed in poultry dressing plants?
9. Name two poultry products which have potential as being the vehicle for the growth of food poisoning bacteria.
10. How can the growth of bacteria on fresh poultry be controlled?

REFERENCES

CENTER FOR DISEASE CONTROL. 1974. Foodborne and waterborne disease outbreaks. U.S. Department of Health, Education, and Welfare.

DAWSON, L. E. and STADELMAN, W. J. 1960. Microorganisms and their control on fresh meats. North Central Regional Publ. *112*, Michigan State Univ.

GUNDERSON, M. F., MCFADDEN, H. W., and KYLE, T. S. 1954. The Bacteriology of Commercial Poultry Processing. Burgess Publishing Co., Minneapolis.

KERELUK, K. and GUNDERSON, M. F. 1960. Survival of bacteria in artificially contaminated frozen meat pies after baking. Appl. Microbiol. *9*, 6–9.

KOTULA, A. and KINNER, J. A. 1964. Airborne microorganisms in poultry processing plants. Appl. Microbiol. *12*, 179–184.

MERCURI, A. J., BANWART, B. J., KINNER, J. A. and SESSOMS, A. R. 1970. Bacteriological examination of commercial precooked eastern-type turkey rolls. Appl. Microbiol. *19*, 768–771.

MOSSEL, D. A. A. and INGRAM, M. 1955. The physiology of the microbial spoilage of foods. J. Appl. Bacteriol. *18*, 232–268.

MORRIS, G. K. and WELLS, J. G. 1970. *Salmonella* contamination in a poultry-processing plant. Appl. Microbiol. *19*, 795–799.

ROMANS, J. R. and ZIEGLER, P. T. 1974. The Meat We Eat. Interstate Printers and Publishers, Danville, Illinois.

SADLER, W. W., YAMAMOTO, R., ADLER, H. E. and STEWART, G. F. 1960. Survey of market poultry for *Salmonella* infection. Appl. Microbiol. *9*, 72–80.

WATT, B. K. and MERRILL, A. L. 1963. Composition of foods. USDA Agri. Handbook *8*.

WEISER, H. H., MOUNTNEY, G. J. and GOULD, W. A. 1971. Practical Food Microbiology and Technology, 2nd Edition. AVI Publishing Co., Westport, Conn.

WOODBURN, M. 1964. Incidence of *Salmonella* in dressed broiler-fryer chickens. Appl. Microbiol. *12*, 492–495.

9

Microbiology of Fish and Seafood

FISH AND SEAFOODS AS SUBSTRATES FOR SPOILAGE BACTERIA

pH

Fish spoilage is related to spoilage in red meats and in poultry meats in that the amount of glycogen in the muscle determines the amount of lactic acid produced when there is a drop in pH. Rigor mortis in fish runs concomitant with acid production in the muscle (Amlacher 1961). As we observed with the other meats, a drop in pH has a definite, beneficial bactericidal or bacteriostatic effect. pH is one of the factors influencing microbial growth. There are species differences in the pH of fish tissue. For example, halibut can have a pH of 5.6 with resulting good storage quality (Amlacher 1961).

In general, the pH of fish flesh just after being caught is slightly alkaline (7.05–7.35). After a while, the pH drops to 6.2–6.5; in mammalian meat the pH may drop to as low as 5.5. On balance, the pH of fish flesh is higher than mammalian flesh (Amlacher 1961).

Struggling fish lower their glycogen content rapidly with a change in pH to values as low as 5.53, but if struggling continues, the pH rises to become more alkaline. The less glycogen in the muscle, the shorter the keeping time of the fish. Flesh from exhausted fish will keep only about half as long as that from nonexhausted fish. It has been shown that trawl-caught haddock with fatigued muscles had an alkaline pH compared to penned fish with rested muscles (Amlacher 1961). For the most part, the pH values are in a range that supports rapid microbial growth (Table 9.1).

TABLE 9.1. pH VALUES OF FISH AND SEAFOODS

Food	Min	Avg	Max
Clams	5.9	6.8	7.1
Codfish	6.0	6.0	6.1
Crabmeat	6.3	6.8	7.4
Mackerel	5.9	6.1	6.2
Oysters	6.3	6.4	6.7
Salmon	6.4	6.4	6.5
Sardines	5.7	6.0	6.6
Shrimp	6.4	7.0	7.3
Tuna	5.8	5.9	6.1

Source: American Can Company (1949).

Vitamins

According to Higashi (1961), fish contain thiamin, riboflavin, pyridoxine, B_{12}, niacin, pantothenic acid, folic acid, and choline. These B complex vitamins may be required by some food spoilage bacteria. The supply of vitamins is adequate for microbial growth.

Minerals

Causeret (1962) found the following minerals in fish: sulfur, chlorine, sodium, potassium, phosphorus, calcium, magnesium, iron, copper, iodine, fluorine, manganese, zinc, lead and arsenic. With the exception of lead and arsenic, these elements are needed in the nutrition of bacteria.

Carbon and Nitrogen Sources

The protein, fat and carbohydrate content of selected fish and seafood are indicated in Table 9.2. As we have discussed previously, the spoilage bacteria can obtain both carbon and nitrogen from the protein of the flesh. The bacterial enzymes can cause breakdown of the fish lipids. The carbo-

TABLE 9.2. PROTEIN, FAT AND CARBOHYDRATE CONTENT OF SELECTED FISH AND SEAFOODS

Food	Protein (%)	Fat (%)	Carbohydrate (%)
Tuna, bluefish—raw	25.2	4.1	0
Salmon, Atlantic—raw, whole	14.6	8.7	0
Sardines, Pacific—raw	19.2	8.5	0
Oysters—raw, meat only	8.4	1.8	3.4
Crabs	17.3	1.9	0.5
Shrimp—raw, fresh only	18.1	0.8	1.5
Lobster—raw, meat only	16.9	1.8	0.5

Source: Watt and Merrill (1963).

hydrate is low or does not exist with the exception of oysters where the spoilage bacteria may change the carbohydrate into lactic acid, causing a drop in pH. This drop in pH is an indicator of bacterial spoilage.

SPOILAGE FLORA

Mossel and Ingram (1955) listed *Pseudomonas, Flavobacterium,* and *Micrococcus* as the genera dominating when spoilage of fish occurs during standard conditions. Shewan (1961) also found the *Pseudomonas-Achromobacter* group predominating during spoilage of fish.

The Gram-negative rods (*Pseudomonas* and *Flavobacterium*) and *Micrococcus* are also dominant in the spoilage of shrimp and shellfish (Mossel and Ingram 1955).

FISHING AND PROCESSING METHOD

Types of Fish and Seafoods and How They Are Caught

A brief discussion of the methods of fishing and the type of marine products caught is necessary for our understanding of the microbiological problems.

Surface-Dwelling Fish. Since tuna, salmon, sardines, mackerel and anchovies live on or near the surface of the sea, they are harvested by purse sieves, gill nets, trolling lures, and hooks and lines.

Bottom-Dwelling Fish and Seafoods. Sole, flounder, cod, haddock, ocean perch, snapper, whiting, shrimp, crab, oysters, and clams are caught by otter trawls, submerged gill nets, hooks and lines, pot traps, dredges and tongs (Dassow and McNeely 1963).

Handling and Preserving

All fish and seafoods should be processed as rapidly as possible. Modern trawlers are both fishing and processing plants. Because the United States does not have many of these ships, our discussion will center on methods of less-developed technologies.

Deck Layout. There should be ample space on the deck for dumping, sorting, dressing and preparing the fish for the hold. Adequate measures for keeping the deck in sanitary condition are necessary. With wooden decks and/or preparation areas, this may be difficult to achieve.

Hold Layout. The hold area where the fish are to be stored should be divided into bins with shelves so that the catch can be iced without being crushed by the combined weight of other fish and ice. One ton of ice is recommended per two tons of fish (Dassow and McNeely 1963). The shelves should be built so that the water is drained away from the fish.

On short fishing trips, the catch is iced for 5 to 15 days. On longer trips, the fish may be refrigerated by sea water. Crabs and lobsters are held in live wells.

After the fishing ships return to port, the fish are washed, iced and packed in wooden boxes for shipping to wholesale distribution points.

Fish to be processed for other means of preservation are washed in chlorinated water, filleted, and frozen. These fillets may be cut into sticks, breaded and frozen. Tuna and salmon are usually canned.

General Factors Affecting the Quality of Fish

1. Because of the rapidity at which fish spoil, all fish should be handled rapidly.

2. The temperature should be maintained as low as possible during handling and processing.

3. Nonporous equipment is recommended for handling fish for ease in maintaining sanitary conditions.

4. At the processing plant on shore, the building should be screened and rodent-proof. An active insect and rodent control program should be in effect.

5. Walls, ceilings and floors should be easy to clean and to keep sanitary.

6. Employees should be instructed on good health and sanitation practices.

7. Adequate clean toilet and handwashing facilities should be available.

8. Chlorinated water should be used in the plant on the equipment and on the product (2–5 ppm on equipment and fish and 10–20 ppm during clean up).

9. Eviscerated and filleted fish should be kept refrigerated and/or frozen until consumed.

Many of these factors also apply to other foods and not necessarily only to fish.

Specific Factors Affecting Microbial Content of Sea-Water Fish

The following list was made from Shewan's chapter (1961) on "The Microbiology of Sea-Water Fish."

(1) At the time fish are caught, the gill tissue and intestinal fluid carry from 10^3 to 10^7 viable microorganisms per sq cm, per g, or per ml.

(2) The flesh of newly-caught healthy fish is generally considered sterile. The surface contains bacteria.

(3) Fish from warmer seas (such as off the African, Indian and Australian coasts) have more mesophiles than do those from colder seas.

(4) Bacterial numbers in the sea are related to the plankton populations so that concentrations of bacteria in the sea vary with the seasons.

(5) Different species of fish and the method used in catching affect the bacteria load. For example, trawled fish have 10 to 100 times more bacteria than lined fish.

(6) If the fish remain on deck very long, there may be a very large increase in bacteria content on the surface of the fish. Deck boards and holding pen boards may increase the counts of the fish as they can be a source of inoculum.

(7) If fish are not gutted immediately, or as soon as possible, autolytic enzymes of the fish will digest the walls of the belly of the fish. It takes several days for the bacteria in the viscera to invade the muscle.

(8) Careful washing of fish can reduce the bacterial load 80–90%.

(9) Trawler ice may become contaminated with bacteria (due to contact with contaminated surfaces) and add bacteria to the fish. Unused ice in the trawler's hold may contain 10^5 to 10^6 per ml.

(10) After icing or refrigeration, there is a change in the makeup of the flora. At first, at 20°C, coryneforms and flavobacteria predominate. *Pseudomonas* grows so readily at 0°C that after 9 to 10 days, population can increase to 10^7 to 10^8 per g. In 10–12 days, the flora includes 60–90% *Pseudomonas*, with the remaining genera being *Achromobacter* and *Flavobacterium*. Thus, *Pseudomonas* becomes the dominant spoilage bacterium. In fresh fish, *P. fragi* (acid-producing pseudomonad) represents only a few percent of the population. In 12–15 days of ice storage, this species accounts for 20–50% of the total count.

SENSORY EVALUATION AND CHEMICAL DETECTION OF SPOILAGE OF FISH

In Freedman's *Sanitary Inspector's Manual* (1942), the characteristics of fresh and stale fish are compared. These comparisons are listed in Table 9.3.

Fish decompose along ribs and backbone. The bacteria causing the spoilage are proteolytic and produce several breakdown products of proteins. A discussion of metabolic by-products of microorganisms as indicators of food quality has been reviewed by Fields *et al.* (1968). Farber (1963) found that there was a correlation of trimethylamine (TMA) nitrogen and sensory judgment for white-fleshed fish but not for spoilage of herring.

TABLE 9.3.   CHARACTERISTICS OF FRESH AND STALE FISH

Fresh	Stale
Bright in appearance as if still alive.	Dull, lifeless in appearance.
Eyes are bright and full.	Eyes are sunken.
Gills are bright red in color.	Gills are dirty in color.
Flesh is firm to the touch.	Flesh is soft and limp.
Abdomen is clean and free of offensive odor.	Abdomen is discolored and odor offensive.
Blood is fresh red in color and of normal consistency.	Blood is dark, thin in consistency and offensive.
Flesh adheres to the bones when split.	Flesh leaves bones easily when split.
Backbone is a pearly gray color.	Backbone is a pink discoloration.

Source: Freedman (1942).

PROCESSING SEAFOODS

Shrimp

Shrimp is the most important seafood in value (Stansby 1963). There are three species of shrimp: brown, white, and pink. For processing, they are sold as frozen with shell on and headless. In more recent years, this seafood has been breaded and then frozen.

In a comprehensive review of the microbiology of shellfish by Fieger and Novac (1961), they state that shellfish spoil more readily than fish because the former contain more free amino acids. Spoilage starts immediately through marine bacteria on the surface of the shrimp as soon as they die. Bacterial counts vary with the place they are caught and with the amount of contamination from the bottom mud.

According to Fieger and Novak (1961), bacterial counts on shrimp from the Gulf of Mexico (off the Louisiana coast) ranged from 1600 to 160,000 per g. They also stated that removal of the head reduced the degree of bacterial contamination by 75%. Relatively low bacterial counts were obtained for headless shrimp which were washed and iced rapidly. As with fish in the holds, bottom layers of shrimp have higher counts than do the upper layers.

Fieger and Novak (1961) observed a constant increase in *Achromobacter* sp. (probably these are *Pseudomonas; Achromobacter* is not recognized as a genus at present) until after 16 days in crushed ice. They comprised 82% of the population.

Shrimp on the market were graded as good, fair and poor quality on the basis of total counts. Good shrimp contained 4.5×10^6 microorganisms per gram; fair, 10.7×10^6; and poor, 19×10^6. On the basis of sensory evaluation, shrimp were divided into five classes. Class 0 had an odor of fresh caught shrimp with no indole present. Class 1 shrimp contained 8.0 mg of indole and had "strong," "old" or "fish" odors. Class 2 with 4.0 mg of indole had a faint, recognizable odor of decomposition. Class 3 (159–161 mg of indole) exhibited a repugnant, deep-seated odor of decomposition. The last class (Class 4) had 992–1080 micrograms of indole and a deep-seated, nauseating odor of putridity (Duggan and Strasburger 1946).

In addition to the breakdown of tryptophan to indole, there is also a slower ammoniacal type of decomposition with the characteristic odor of free ammonia.

Fieger and Novak (1961) stated that there was a definite correlation with taste panel scores, logarithm of total bacterial count, and the ratio of volatile bases (TVB) to total nitrogen (TN) multiplied by 100 or TVB/TN × 100. On the basis of these ratios and the average taste panel scores, four groups for determining quality of breaded shrimp were suggested. Ratio values below 6.0 indicated very good quality; those between 6.0 and 7.5 were rated as good; values between 7.5 and 8.5 were listed as questionable; and values higher than 8.5 were considered to be of unacceptable quality.

It is understood that shrimp should not be taken from water polluted by human feces.

Crabs

Crabs may be marketed either alive or immediately after cooking. Crabs to be processed are butchered (viscera and gills are removed), and then cooked 12–15 min in steam. After cooling, the crab meat is removed. It takes 100 lb of blue crabs to make 10 to 18 lb of crab meat. Pasteurizing the meat at 76.6°C (170°F) for about 1 min prior to shipping in ice extends the shelf-life. Crab meat may be frozen and/or canned (Stansby 1963). If canned, the crab meat is processed in a No. 2 can at 115.5°C (240°F) for 60 min.

According to Fieger and Novak (1961), *Proteus, Pseudomonas* and *Flavobacterium* spoil crab meat. *Proteus,* which grows rapidly in this food, can cause food poisoning. With growth of spoilage bacteria, there is an increase in pH from 7.2 to 7.8–8.2 at the onset. The pH of fresh crab meat is 7.2–7.4, whereas spoiled meat has a pH of 8.0 to 8.5.

Microorganisms on contaminated fresh crabs can be reduced by placing them in noncontaminated water. Although this method reduces fecal organisms, some are retained. The same procedure is used with oysters.

Ammonia has been used as an index of quality of crabmeat. According to Burnett (1965), fresh meat contained 0 to 0.4 mg per gram; meat which had a fishy odor but was still judged to be good had an average of 14.3 mg per gram; meat with a slight but definite decomposition odor averaged 70.4 mg per gram; meat with advanced decomposition contained 128.5 mg per gram; and putrid meat had values as high as 175.8 mg per gram of meat. There is a definite agreement between sensory evaluation and chemical analyses of the meat.

Oysters

Oysters are grown on the east and west coasts. They are dredged and brought to the processing plant to be shucked, cleaned, graded and packed. After the shells are opened by hand, the oysters are placed in a vat with potable water where they are agitated with compressed air to remove sand.

After being washed, the oysters are graded by hand on tables. Workers must practice sanitation for a good product. Because oysters may be consumed raw, not only should the plant and processing conditions be sanitary, but also the oysters should be grown in nonpolluted water (Stansby 1963).

Spoilage flora include: *Serratia, Pseudomonas, Proteus, Clostridium, Bacillus, Enterobacter,* and *Escherichia.* These bacteria initiate spoilage. Good oysters have a pH of 6.2 to 5.9; oysters with a pH of 5.8 are classified as "off"; when the pH drops to 5.7 to 5.5, they are classified as "musty"; and when the pH drops to 5.2, they are sour or putrid (Fieger and Novak 1961).

With the drop in pH, conditions now favor the more acid-tolerant microorganisms. Hence, the latter stages of spoilage are due to streptococci, lactobacilli and yeasts.

The degree of washing of the shucked oysters and the quality of water influence the course of decomposition. Clean water removes the putrifying agents but favors fermentation and souring in the ensuing deterioration.

Beacham (1946) tested the relation between indole content of oysters and sensory characteristics. Sound oysters had no odor and only 1.2 to 3.0 mg of indole per 100 g; oysters with a lightly "off" odor contained 16.5 to 26.6 mg of indole per 100 g; oysters with a repugnant odor had 53.0 to 87.2 mg of indole per 100 g. Indole, a breakdown product of tryptophan, gives added evidence of microbial decomposition.

Oysters take on their own weight in water every two hours, so that if there is fecal pollution, they will contain *Escherichia coli.*

Enterobacter cloacae which is part of the normal flora of oysters will give a positive presumptive and confirmed test for *E. coli. E. cloacae* is a coliform intermediate (a form which has characteristics between *Enterobacter aerogenes* and *E. coli*). However, if the sample and lactose broth are incubated at 45°C, *E. coli* will grow but *E. cloacae* will not (this is the Eijkman test). Emphasis must be placed upon the presence of *E. coli,* not coliforms in general.

In October, 1975 in the FDA Consumer, Emil Corwin wrote an article on "For Safer Shellfish." This article is reprinted below in its entirety.

For Safer Shellfish

Because of where they live, how they eat, and how they are eaten, oysters, clams, and mussels present some special sanitation problems. To assure a continuing supply of safe and wholesome shellfish, FDA has proposed new regulations to strengthen the 50-year-old National Shellfish Sanitation Program.

Fifty years ago, following a typhoid fever outbreak traced to the consumption of oysters from contaminated waters, the Federal Government, the States, and the shellfish industry agreed to cooperate to improve sanitary practices to ensure that shellfish such as oysters, clams and mussels are safe, pure, and wholesome.

This agreement led to the formation of the National Shellfish Sanitation Program (NSSP), one of the oldest voluntary tripartite relationships of its kind in the United States. In most instances, the program has succeeded in its primary objective—to prevent the harvest and interstate shipment of bivalve molluscan shellfish from contaminated waters, and prevent shellfish from becoming contaminated during processing. Despite the high consumption of shellfish in the United States, there have been relatively few occurrences of shellfish-borne infectious diseases.

Why, then, did the Food and Drug Administration on June 19, 1975, find it necessary to propose new regulations to strengthen the NSSP? Part of the answer lies in the nature of bivalve shellfish and the way they are processed and consumed. Among the factors that complicate shellfish sanitation are:

1. In the process of feeding, oysters, clams, and mussels filter and retain harmful organisms and toxic substances present in the water in which they live. High concentrations of these substances may be present in shellfish meat.

2. The environment in which shellfish grow is almost always subject to some degree of human, industrial, or animal pollution.

3. Shellfish often are packed whole and alive and often are consumed either raw or partially cooked.

4. Extraneous substances can be introduced into shellfish meats during the shucking process.

5. Raw shellfish packed and held under inadequate refrigeration can provide excellent growth media for bacteria.

Moreover, in addition to contamination introduced into the environment by man, shellfish may become toxic from natural causes. Under certain environmental conditions, large masses of marine algae, some of which produce a toxin, will occur in the sea. Shellfish, through their normal feeding mechanism, ingest such algae and the toxin becomes concentrated in their tissue. Persons who eat these toxic shellfish may become temporarily paralyzed and, in very severe cases, die. Toxic shellfish are associated with three species of marine algae that bloom sporadically along the west coast from Alaska to California, along the east coast of Maine to Massachusetts, and along the Florida Gulf Coast.

Given the vulnerability of shellfish to contamination and the highly perishable nature of shellfish products, careful and constant control of harvesting and processing procedures must be maintained, and regulations strictly enforced, if the public is to continue to have available safe and wholesome oysters, clams, and mussels. It has become increasingly clear that to assure such control and enforcement the NSSP must be strengthened. Two recent surveys, one by the U.S. General Accounting Office, an arm of the Congress, and another by FDA, showed that some States were not adequately inspecting the waters from which shellfish are harvested or the way shellfish are processed for marketing. Another problem pointed up by the surveys was the inconsistency in procedures for dealing with interstate producers or processors who violate shellfish sanitation regulations.

The regulatory changes proposed by FDA are intended to correct these shortcomings and to make other improvements in shellfish sanitation. The proposed regulations would, for the first time, establish nationally uniform administrative procedures under which the 50-year-old NSSP would operate. The procedures would define the scope, requirements, and responsibilities of the State and Federal Government agencies involved.

More specifically, the proposed regulations would:

1. Require the States to develop a Comprehensive State Shellfish Control Plan (CSSCP) under which they would inform FDA of the measures they will take to inspect shellfish harvesting and processing operations and of the resources they will provide to carry out such surveillance. The CSSCP would also require States to report immediately any conditions involving shellfish that pose a hazard to human health. Under the present system there are no clearly defined procedures for FDA to deal with interstate shellfish producers or products that are found to be in violation of sanitation regulations. The new control plan—a major innovation in shellfish regulation—would enable FDA to take direct action against violative plants when States fail to do so.

2. Require strict enforcement of the present microbiological, pollution, and other quality standards for the Nation's shellfish-growing waters. A 1973 FDA survey of the 22 shellfish-producing States' programs showed that seven States did not meet present NSSP standards for a significant number of their growing areas, yet were allowing harvesting, contrary to control measures stipulated in the NSSP Manual of Recommended Practices.

3. Require a tagging and record keeping system for shellfish that would enable FDA or State authorities to (a) determine within a matter of hours where shellfish that are found to be contaminated were grown; and (b) provide authorities with a definitive list of all producers and processors who may have handled them. The present NSSP labeling system is not being uniformly applied, depriving officials of a rapid and reliable method of tracing contaminated shellfish to their growing area.

4. Establish specific control practices and sanitary requirements for both processors of shellfish and handlers of shell stock (unshucked products). Enforcement of requirements for water quality standards and growing area identification systems is not enough to ensure that only sanitary shellfish reach the consumer. An effective shellfish sanitation program must include controls at all stages in the operation. As with water quality standards, the States would continue to bear primary responsibility for enforcing requirements for processing and handling, but FDA would also be prepared to take appropriate steps where necessary.

5. Require imported fresh or frozen shellfish to meet the same sanitary standards as those applied to shellfish produced in the United States. Im-

ported shellfish comprise almost 10 percent of the shellfish in these categories consumed in this country.

Although controls are required at all stages of shellfish harvesting and processing, it is the quality of the water in which shellfish are grown that is of primary concern. Shellfish pump vast quantities of water through their bodies as an essential part of their life processes. Some species of oyster, for instance, pump as much as nine gallons of water per hour for hours at a time. It is in this pumping process that shellfish may accumulate and concentrate harmful microorganisms, toxic chemicals, and heavy metals from their marine environment. If the shellfish are exposed to water polluted with human and animal excrement, they become agents for disease.

Sewage-contaminated oysters caused the widespread typhoid fever outbreak in the winter of 1924 that led to the shellfish sanitation program. There were cases in cities as far apart as New York, Chicago, and Washington, D.C. Local and State public health officials and the shellfish industry became sufficiently alarmed to request that the Surgeon General of the United States Public Health Service develop the necessary control measures to ensure safe and wholesome shellfish.

What has happened since provides an interesting insight into the evolution of a consumer protection program revised over the years to meet changing conditions and new needs.

The Surgeon General called a conference that included representatives from State and municipal health authorities, the Bureau of Chemistry (later to become the FDA), the Bureau of Fisheries (now the National Marine Fisheries Service), and the shellfish industry. This historic conference in Washington, D.C., on February 19, 1925, decided that the States should bear full responsibility for the sanitary control of their own shellfish resources and that the Public Health Service should serve as a clearinghouse for information on the effectiveness of State control programs. The Public Health Service's role was later changed from fact-gathering to appraising each State's sanitation program and that agency was given authority to endorse or refuse endorsement of any program.

After more than four decades under the Public Health Service, Federal jurisdiction of administering the NSSP was transferred to the FDA in 1968. This meant that FDA was now authorized under the Public Health Service Act as well as the Food, Drug, and Cosmetic Act to improve shellfish control programs. In meeting its new responsibilities, FDA not only continued to review and appraise State programs, but initiated a number of new measures. The Agency drafted a good manufacturing practice (GMP) regulation to guide shellfish processors and handlers, trained its staff in techniques for inspecting shellfish-producing areas, established closer liaison with State shellfish control agencies, and added facilities and conducted training for State, Federal, and foreign shellfish sanitation experts.

FDA also has expanded shellfish research. It conducts research at its Gulf Coast laboratory at Dauphin Island, Alabama, mainly in shellfish purification processes (depuration), and supports research elsewhere in problems related to shellfish contamination. The Agency has budgeted $300,000 this fiscal year for contracts with outside experts to study the intake by shellfish of such contaminants as viruses, bacteria, heavy metals, parasites, and paralytic shellfish poison.

FDA has sought the advice of the States and of shellfish producers and processors in its efforts to strengthen the NSSP and to make it a model Federal-State-Industry food protection program. For the past two years, the Agency has given careful attention to ideas and suggestions proposed at national shellfish workshops and regional meetings and has studied the many comments received in response to a prepublication draft of the proposed regulations.

It is from this background that the present proposed regulations have emerged. They represent a unique blend of State and Federal authorities to achieve a common objective—safe and wholesome shellfish. In seeking to strengthen the existing NSSP, the regulations establish procedures which, for the first time, would enable FDA to act promptly in emergency situations, including authority to quarantine harvesting areas to prevent contaminated products from entering interstate commerce.

Since the new regulations would encompass all aspects of safety, not merely sanitation, FDA has recommended that they be known collectively as the "National Shellfish Safety Program." It's still NSSP, but the initials would mean a lot more than they did a half century ago.

GLOSSARY

Autolytic enzymes—Enzymes contained in food itself as contrasted to enzymes from microorganisms.

Bactericidal—To kill bacteria.

Bacteriostatic—To arrest or inhibit bacterial growth.

Confirmed test—One of the three tests (presumptive, confirmed and completed) for *E. coli.*

Fillet—To remove a strip or compact piece of boneless fish.

Hold—The interior of a ship, below decks, where the fish or seafoods are stored.

Indole—A breakdown product from the amino acid tryptophan which is used as a chemical indicator of quality.

Presumptive test—See confirmed test.

Rigor mortis—Muscular stiffening following death.

Sensory evaluation—Evaluation made by using one or more of the senses.

Shellfish—Aquatic animals having a shell or shell-like exoskeleton.

Shucked oysters—Oysters with the outer shell removed.

Trawler—A boat used for trawling.

Trawler ice—Ice stored aboard the trawler to cool the seafoods.

Trawling—To catch seafood by towing a large, tapered net of flattened conical shape along the sea bottom.

REVIEW QUESTIONS

1. Compare and contrast the glycogen present in muscle of fish, beef, and poultry on pH of the meat after slaughter. What influence does this have on spoilage?
2. List the pH of fish and seafoods.
3. Discuss fish and seafoods as substrates for bacterial growth.
4. Do bacteria need lipids for growth?
5. What seafood has the most carbohydrate?
6. Why do seafoods need to be grown in nonpolluted water?
7. What is the dominant flora which spoils fish under standard conditions?
8. List methods of harvesting fish. Which methods may contribute to the flora?
9. What is the ideal way of processing fish? What are some of our methods?
10. What are the factors contributing to high bacteria counts on board ship?
11. What genus and species dominate in numbers on fish after a considerable storage time in crushed ice?
12. Compare and contrast characteristics of fresh and stale fish.
13. Which seafood has the most value?
14. List some metabolic by-products used to judge the quality of fish and other seafoods.
15. What effect does beheading shrimp have on the bacterial count of shrimp?
16. What is the chemical compound which results from bacterial breakdown of tryptophan?
17. How may crab meat be processed?
18. List bacteria spoiling crab meat.
19. What metabolic by-product has been used to judge quality of crab meat?
20. List the flora spoiling oysters.
21. What indicator bacterium is used to determine whether or not there is fecal pollution of oysters?
22. List the succession of microorganisms which spoil oysters.
23. What bacterium can give a false presumptive test when oysters are assayed for fecal pollution?
24. What is the Eijkman test?
25. What metabolic by-products have been used to judge the quality of oysters?

REFERENCES

AMERICAN CAN COMPANY. 1949. The Canned Food Reference Manual. American Can Co., New York.

AMLACHER, E. 1961. Rigor mortis in fish. *In* Fish as Food, Production, Biochemistry, and Microbiology, Vol. 1. Georg Borgstrom (Editor). Academic Press, New York.

BEACHAM, L. M. 1946. A study of decomposition in canned oysters and clams. J. Assoc. Offic. Agr. Chemists *29*, 89–99.

BURNETT, J. L. 1965. Ammonia as an index of decomposition in crabmeat. J. Assoc. Offic. Agr. Chemists *48*, 624–627.

CAUSERET, J. 1962. Fish as a source of mineral nutrition. *In* Fish as Food, Nutrition, Sanitation and Utilization, Vol. 2. Georg Borgstrom (Editor). Academic Press, New York.

CORWIN, E. 1975. For safer shellfish. FDA Consumer *9*, 9–11.

DASSOW, J. A. and MCNEELY, R. L. 1963. Commercial fishery methods. *In* Food Processing Operations, Their Management Machines, Materials, and Methods, Vol. 1. M. A. Joslyn and J. L. Heid (Editors). AVI Publishing Co., Westport, Conn.

DUGGAN, R. E., and STRASBURGER, L. W. 1946. Indole in shrimp. J. Assoc. Offic. Agr. Chemists *29*, 177–188.

FARBER, L. 1963. Quality evaluation studies of fish and shellfish from certain northern European waters. Food Technol. *17*, 476–480.

FIEGER, E. A. and NOVAK, A. F. 1961. Microbiology of shellfish deterioration. *In* Fish as Food, Production, Biochemistry and Microbiology, Vol. 1. Georg Borgstrom (Editor). Academic Press, New York.

FIELDS, M. L., RICHMOND, B. S., and BALDWIN, R. E. 1968. Food quality as determined by metabolic by-products of microorganisms. *In* Advances in Food Research. C. O. Chichester, E. M. Mrak and G. F. Stewart (Editors). Academic Press, New York.

FREEDMAN, B. 1942. Sanitary Inspector's Manual. Louisiana State Department of Health (Printed by Peerless Printing Co., New Orleans).

HIGASHI, H. 1961. Vitamins in fish—with special reference to edible parts. *In* Fish as Food, Production, Biochemistry and Microbiology, Vol. 1. Georg Borgstrom (Editor). Academic Press, New York.

MOSSEL, D. A. A. and INGRAM, M. 1955. The physiology of the microbial spoilage of foods. J. Appl. Bacteriol. *18*, 233–268.

SHEWAN, J. M. 1961. The microbiology of sea-water fish. *In* Fish as Food, Production, Biochemistry and Microbiology. Georg Borgstrom (Editor). Academic Press, New York.

STANSBY, M. E. 1963. Processing of seafoods. *In* Food Processing Operations, Their Management, Machines, Materials, and Methods, Vol. 1. M. A. Joslyn and J. L. Heid (Editors). AVI Publishing Co., Westport, Conn.

WATT, B. K. and MERRILL, A. L. 1963. Composition of foods. USDA Agr. Handbook, *8*.

10

Microbiology of Eggs and Egg Products

EGGS AS A SUBSTRATE FOR MICROBIAL GROWTH

pH

The average pH of newly-laid eggs is 7.6 to 7.9 (Powrie 1973). However, the pH of the albumen may reach 9.7 as a maximum value. A high pH (9.7) would tend to prevent the growth of spoilage bacteria. This rise is due to loss of carbon dioxide. This increase in alkalinity may be prevented by storing shell eggs in an environment of carbon dioxide; however, the pH would then remain in a range for optimum microbial growth.

The pH of the yolk of just-laid eggs is near 6.0 and rises only slightly during storage. The pH values of 6.4 to 6.9 may be reached (Powrie 1973).

Vitamins

Cook and Briggs (1973) found B complex vitamins in whole eggs, yolks and whites. From the standpoint of microbial growth, the vitamin content is adequate.

Minerals

The ash content of chicken eggs is given in Table 10.1. According to Watt and Merrill (1963), the following minerals are components of eggs: calcium, phosphorus, iron, sodium and potassium.

TABLE 10.1.   COMPOSITION OF CHICKEN EGGS

Component	Water Content (%)	Protein (%)	Fat (%)	Carbohydrate (%)	Ash (%)
Whole	73.7	12.9	11.5	0.9	1.0
Whites	87.6	10.9	Trace	0.8	0.7
Yolks	51.1	16.0	30.6	0.6	1.7

Source: Watt and Merrill (1963).

Carbon-Nitrogen Source

Carbon can be obtained from protein and/or carbohydrate. Nitrogen is obtained, of course, from protein. Again, fat is not needed for microbial growth per se, but the microflora can bring about changes in the lipid content.

Inhibitors

According to Board (1973), the albumen of a hen's eggs contains the following factors which may influence microbial growth:

(1) Lysozyme. This is an enzyme which will lyse cell walls of Gram-positive bacteria.
(2) Conalbumin chelates iron, copper and zinc. This may prevent the microorganism from obtaining these minerals.
(3) Avidin combines with biotin to make it unavailable to microorganisms requiring this vitamin.
(4) Riboflavin (a B complex vitamin) chelates cations. This, again, may limit the growth of some of the microflora.
(5) Protein B inhibits fungal protease. The inhibition of proteolytic enzymes, of course, limits the growth of these microorganisms.

As far as nutritional value to man, the egg is the most perfect protein. All other proteins have protein efficiency ratios less than that of egg. The inhibitors and pH tend to limit the growth and types of spoilage flora.

THE SHELL AS A BARRIER

The shell is about $\frac{1}{60}$ of an inch thick and contains pores. The pores are filled with a protein-like substance which can be digested by enzymes of bacteria and molds.

A recently laid egg has a characteristic appearance because of an extremely thin layer of protein material deposited on the shell just before the egg leaves the hen. The shell is composed of $CaCO_3$, 93.7%; $MgCO_3$, 1.39%; P_2O_5, 0.76%; and organic matter, 4.15%.

Under the shell is the shell membrane. According to Board (1973), its exact role in the egg's defense is still uncertain. However, spoilage bacteria grow very well on shell membranes added to mineral salts.

SPOILAGE FLORA

According to Mossel and Ingram (1955), the dominant spoilage flora under standard conditions of storage are *Pseudomonas, Cladosporium, Penicillium* and *Sporotrichum.* Board (1973) listed *Pseudomonas, Alcaligenes, Proteus,* and *Escherichia* as bacteria commonly occurring in rotten eggs. Molds are less important than bacteria if the shell eggs are stored properly, according to Board. Thus, there appears to be some disagreement between Mossel and Ingram (1955) and Board (1973) since three of the four microorganisms listed by Mossel and Ingram are molds.

STORAGE OF SHELL EGGS

Candling

A light source is used to reveal the contents of the egg. Defects, rots, fertile eggs, etc., show up and can be eliminated before storage.

Egg Washing

Washing eggs before storage may increase the chance of spoilage. However, if the shell egg is to be processed (whole; whites; yolks; either frozen or dried) unbroken eggs must be washed to remove manure from the surface of the shell before being broken. Water at 32.2°–60.0°C (90°–140°F) (depending upon the chemicals) is used, usually with a detergent and hypochlorite compound.

Stabilization or Thermostabilization

A heat treatment of 54.4°C (130°F) for 15 min has been recommended by the Poultry Husbandry Department of the University of Missouri. The following are benefits of such a treatment:

(1) Heat devitalizes fertile eggs so that no embryonic development is possible.

(2) This process stabilizes the thick albumen so that the change to thin white occurs slowly.

(3) The egg shell is pasteurized. Consequently, many of the spoilage organisms are killed before storage.

Storage Conditions

The freezing point of eggs is −2.2°C (28°F). Shell eggs should not be stored lower than −2.2°C (28°F). Temperatures at −1.7° to −0.6°C (29°–31°F) may be used at relative humidity of 80–92%. Ozone at a concentration of 1.5 ppm is recommended. Fluctuations in temperature should be avoided to prevent moisture from condensing on shells. A mycostat on crates, flats and fillers will inhibit mold growth (Desrosier and Desrosier 1977).

EGG-BREAKING PLANTS

Prior to freezing or dehydration, eggs are candled to eliminate rots. Eggs should be cleaned (washed with detergents and chlorinated) to reduce any contamination of the processed product.

All equipment in the plant should be made of stainless steel and kept in a sanitary condition. Prior to freezing or dehydration, the whole eggs, egg whites or egg yolks are pasteurized to eliminate salmonellae. After pasteurization, the egg products should be handled very carefully to prevent recontamination.

PASTEURIZATION OF EGG PRODUCTS

Because of the possibility of food infections by salmonellae, USDA now requires that liquid whole egg be heated to at least 60°C (140°F) and held for no less than 3.0 min for the average particle as it passes through a tubular heat exchanger. To achieve the same degree of pasteurization in 3.5 min when salt is added, the temperature is raised to 63.3°C (146°F). Heat treatment at 63.3°C (146°F) for 3.5 min is also required when sugar is added to yolks. The white of the egg with its higher pH and fewer ingredients to protect the bacterial cells needs less heat to achieve the same degree of pasteurization. Table 10.2 summarizes temperature and holding time requirements for liquid egg products.

TABLE 10.2.   MINIMUM TEMPERATURES AND HOLDING TIMES FOR THE AVERAGE PARTICLE

Liquid Egg Product	Minimum Temp Required °C	°F	Minimum Holding Time in Minutes
Whole egg	60.0	140	3.0
Whole egg with salt added (2% or more)	63.3	146	3.5
Whole egg with sugar added (2–12%)	61.1	142	3.5
Plain yolk	61.6	142	3.5
Yolk with sugar added (2% or more)	63.3	146	3.5
Yolk with salt added (2–12%)	63.3	146	3.5
White (no chemicals added)	56.7	134	3.5

Source: Federal Register (12/23/69).

FROZEN EGG PRODUCTS

The freezing process kills some of the bacterial population which may have survived pasteurization. *Bacillus,* because of its spores, is the major type found in the bacterial flora. *Alcaligenes* and *Proteus* can also survive both pasteurization and freezing. Shafi *et al.* (1970) found psychrophilic, thermophilic and anaerobic bacteria in commercial egg products which had been pasteurized and frozen.

DEHYDRATED EGG PRODUCTS

Some bacteria are sensitive to the heat during drying and will die. Others will not. Any *Bacillus* spp. present in the spore state will not be killed. If the egg product to be dehydrated is pasteurized prior to drying, the total population will be reduced but still leave more heat-resistant non-spore-formers and sporeforming bacteria.

If the carbohydrate of egg white is removed, the egg white can be heated to higher temperatures than if the sugar were still present. Heating egg white with glucose present causes the product to brown. Egg white can also be heated after drying to reduce bacterial populations. *Salmonella* subjected to heat treatments in the liquid state (pasteurization) prior to drying and also after drying is more susceptible to the heat treatment (Bergquist 1973).

MICROBIOLOGICAL ANALYSES AND INTERPRETATION OF THE RESULTS (AMERICAN PUBLIC HEALTH ASSOCIATION 1958)

A. Some bacteria grow rapidly in broken-out egg products (whole eggs or yolks) if the temperature rises to 15.6°C (60°F) and the product is held for several hours.

B. Low grade eggs generally have high numbers of bacteria on their shells. Products obtained from such eggs will have high counts, even after pasteurization.

C. Poor sanitation in the plant, especially poor cleaning of holding tanks, pipe lines and pumps, causes high counts in the finished product.

D. If poor quality eggs are used or the eggs are not washed properly, there may be high numbers of bacteria in the finished products.

E. Freezing storage of dried eggs results in marked decrease in numbers of viable bacteria high in numbers initially. Pasteurization lowers the count. Therefore, low counts do not necessarily mean a high-quality product.

F. Since glucose naturally present in egg whites may cause a dark product, egg albumen is fermented to remove the sugar. This fermentation will result in high numbers of bacteria. Since nonfecal forms of coliforms (pure cultures) may be used for the removal of glucose, the product will have high coliform counts. Bad odors in the egg white indicate an uncontrolled fermentation.

G. Direct microscopic counts are used to indicate the type of shell eggs and the sanitary conditions used in breaking them. This count reveals both dead and living cells. Abnormal microscopic counts are indicative of poor quality regardless of the viable count.

H. The presence of mold hyphae and large bacilli indicates a product made from shell eggs that have been held in storage. Many short diplocells with pointed ends indicate poor sanitation and handling conditions in the egg-breaking plant.

CHEMICAL INDICATORS OF QUALITY OF EGG PRODUCTS

Formic, acetic, lactic and succinic acids have been suggested as chemical indicators of quality for egg products (Fields *et al.* 1968). For frozen and liquid egg products a direct microscopic count (DMC) greater than 5,000,000 per g and a lactic acid content greater than 7 mg per 100 g are indicative of decomposition. Higher values are given for dried products (DMC greater than 100,000,000 per g and lactic acid content greater than 50 mg/100 g) (Lepper *et al.* 1944).

GLOSSARY

Albumen—A nutritive substance surrounding a developing embryo, as the white of a hen's egg.

Albumin—Several simple water-soluble proteins that are coagulated by heat and are found in egg white.

Avidin—A protein in egg albumin that inactivates biotin.

Candling—A process in which eggs are placed in front of a light source in a dark environment to view the internal contents.

Conalbumen—A glycoprotein found in egg white which complexes with iron, zinc, and copper.

Stabilization or thermostabilization—A mild heat treatment to kill a fertile egg without coagulating the proteins.

REVIEW QUESTIONS

1. Describe the hen egg as a substrate for bacterial growth.
2. What inhibitors are found in eggs?
3. How much of a barrier is the shell to spoilage microorganisms?
4. List spoilage flora of eggs.
5. What should be done to protect shell eggs during storage?
6. Describe pasteurization of egg products.
7. What effect do freezing and drying have on the total microbial flora?
8. Name chemical compounds which have been used as chemical indicators of quality in egg products.
9. Discuss the following in egg products:
   (a) High coliform count in egg white.
   (b) A DMC of 5,000,000 per g with a lactic acid content of 7 mg/100 g.
   (c) Many short diplocells with pointed ends in the product.
10. Discuss *Salmonella* in egg products.

REFERENCES

AMERICAN PUBLIC HEALTH ASSOCIATION. 1958. Recommended Methods for the Microbiological Examination of Foods. American Public Health Association, New York.

BERGQUIST, D. H. 1973. Egg dehydration. *In* Egg Science and Technology, 2nd Edition. W. J. Stadelman and O. J. Cotterill (Editors). AVI Publishing Co., Westport, Conn.

BOARD, R. G. 1973. The microbiology of eggs. *In* Egg Science and Technology,

2nd Edition. W. J. Stadelman and O. J. Cotterill (Editors). AVI Publishing Co., Westport, Conn.

COOK, F. and BRIGGS, G. M. 1973. Nutritive value of eggs. *In* Egg Science and Technology, 2nd Edition. W. J. Stadelman, and O. J. Cotterill (Editors). AVI Publishing Co., Westport, Conn.

DESROSIER, N. W. and DESROSIER, J. N. 1977. The Technology of Food Preservation, 4th Edition. AVI Publishing Co., Westport, Conn.

FIELDS, M. L., RICHMOND, B. S. and BALDWIN, R. E. 1968. Food quality as determined by metabolic by-products of microorganisms. *In* Advances In Food Research. C. O. Chichester, E. M. Mrak and G. F. Stewart (Editors). Academic Press, New York.

LEPPER, H. A., BARTRAM, M. T. and HILLIG, F. 1944. Detection of decomposition in liquid, frozen and dried eggs. J. Assoc. Offic. Agr. Chemists *39*, 185–194.

MOSSEL, D. A. A. and INGRAM, M. 1955. The physiology of the microbial spoilage of foods. J. Appl. Bacteriol. *18*, 233–268.

POWRIE, W. D. 1973. Chemistry of eggs and egg products. *In* Egg Science and Technology, 2nd Edition. W. J. Stadelman and O. J. Cotterill (Editors). AVI Publishing Co., Westport, Conn.

SHAFI, R., COTTERILL, O. J. and NICHOLS, M. L. 1970. Microbial flora of commercially pasteurized egg products. Poultry Sci. *49*, 578–585.

WATT, B. K. and MERRILL, A. L. 1963. Composition of foods. USDA Agr. Handbook *8*.

11

Microbiology of Dehydrated Fruits, Vegetables and Cereals

SUBSTRATE EFFECTS ON GROWTH OF MICROORGANISMS

Water Content

The moisture content of selected fruits and vegetables is given in Table 11.1. Of the latter, lima beans and corn contain the lowest amount of moisture. Such vegetables as peas, white potatoes and sweet potatoes are high in starch content. The total carbohydrate content of vegetables in Table 11.1, excluding vegetables high in starch (lima beans, corn, peas, white and sweet potatoes), is lower on the average than that of the fruits (6.6% vs 12.9%). These vegetables contain less sugar than do fruits. Osmotic pressure is a function of the number of particles in solution. Since fruits contain more sugar, higher osmotic pressure occurs in dried fruit products than in the dried vegetable products. For this reason, dried fruits can retain more moisture without deterioration of quality. (See osmotic pressure in Chap. 5.)

pH

Since the pH of fruits is more acid than that of vegetables, dehydrated fruits are attacked by yeasts and molds. Molds and bacteria grow on dehydrated vegetables. The pH, however, does not have as much influence upon spoilage as does moisture.

TABLE 11.1.   MOISTURE AND CARBOHYDRATE CONTENT OF SELECTED FRUITS AND VEGETABLES

	Moisture (%)	Total CHO (%)	Fiber (%)
Vegetables			
Asparagus[1]	92.7	3.6	0.9
Bean (green)[1]	91.5	5.6	1.0
Lima beans	68.5	21.0	1.0
Beets[1]	87.4	10.0	0.9
Broccoli[1]	89.1	7.4	0.8
Cabbage[1]	93.0	4.4	0.8
Carrots[1]	89.6	8.3	0.9
Celery[1]	92.8	4.2	1.0
Corn[1]	62.5	30.7	1.1
Onions[1]	88.6	9.0	0.7
Peas	75.6	16.9	2.4
Potatoes (white)	78.3	18.7	0.4
Sweet potatoes	72.3	25.6	0.8
Fruits[2]			
Apples	85.8	13.2	0.6
Apricots	88.6	9.0	1.5
Bananas	71.6	26.1	0.6
Cherries	87.4	10.3	1.0
Cranberries	87.9	10.8	1.4
Figs	83.6	14.5	1.5
Grapes (raisins)	86.0	12.8	0.9
Peaches	87.9	10.4	1.8
Pears	87.6	11.4	1.0
Prunes (plum)	87.4	11.4	0.6

Source: Leung *et al.* (1972).
[1] Average CHO of these vegetables: 52.5/9 = 5.8.
[2] Average CHO of all fruits listed: 129.9/10 = 12.9.

Response of Microorganisms to Different Moisture Levels

Available moisture (a_w) may be expressed in terms of water activity as follows:

1. $a_w = \dfrac{\text{Vapor pressure of the solution or food}}{\text{Vapor pressure of water}}$

2. a_w for pure water = 1.0.

3. a_w for 1.0 molal solution of a non-ionizing solute (sugar for example) has an a_w of 0.9823.

4. The a_w would be in equilibrium with a relative humidity of the atmosphere above the food in a closed container. $100 \times a_w$ = equilibrium relative humidity (ERH).

5. An ERH above a food that corresponds to an a_w lower than the food lowers the a_w of the food; or an ERH higher than the food causes the food to pick up moisture from the air. Data in Table 11.2 show the increases in moisture content of vegetables held at an ERH higher than that of the product. According to Mossel and

TABLE 11.2.   EQUILIBRIUM MOISTURE DATA

Product	ERH of Environment	% Moisture (Dry wt) at 37°C
Carrots, Chantenay, unblanched	10	1.8
	20	3.1
	30	4.9
	40	8.0
	50	11.9
	60	17.1
	70	24.2
Cabbage, Savoy, unblanched	10	2.3
	20	3.8
	30	5.2
	40	7.3
	50	10.3
	60	15.1
	70	22.2
Potatoes, Russett, unblanched	10	4.2
	20	6.0
	30	7.4
	40	8.9
	50	10.9
	60	13.2
	70	16.1

Source: Data taken from Makower and Dehority (1943).

Ingram (1955), an ERH of 70% (a_w of 0.70) gives reasonable protection from spoilage. These authors use the term "alarm water content" to mean the highest water content at which microbial spoilage will not occur. The alarm water content based upon an a_w of 0.70 is 15% moisture at 20°C for milk and 14–20% moisture for dehydrated vegetables.

Mossel and Ingram (1955) grouped microorganisms according to the lowest values of ERH/100 or a_w permitting development. These are:

Normal bacteria	0.91
Normal yeasts	0.88
Normal molds	0.80
Halophilic bacteria	0.75
Xerophilic fungi	0.65
Osmophilic yeasts	0.60

Mold Growth on Cereal Grain

Tuite and Christensen (1957) studied the growth of molds on wheat at moisture levels between 12.2 and 16.0% at 25°C (77°F) for 1 to 15 months. The results are given in Table 11.3. Moisture levels are critical in a narrow range from 13 to 15%. Just an increase of 0.5% moisture in the range 14.2

TABLE 11.3.   INVASION OF WHEAT BY MOLDS AT DIFFERENT MOISTURE LEVELS

Moisture Level (%)	Invasion Rate
13.0–13.6	Gradually by *Aspergillus restrictus* and by *A. amstelodami; A. repens* and *A. ruber* more slowly.
14.3–14.6	*A. repens* invaded a larger percentage of seeds in 2 months than other species. After 4 months, all species had invaded the seeds.
15.5–16.0	Complete invasion in 1 month.

Source: Tuite and Christensen (1957).

to 15.5% greatly enhanced the growth of mold during a period of 32 days.

Fungal attack of cereals begins in the field and before the grain is field-dry. Some of these fungi may produce mycotoxins. Molds can be separated into groups, too, on the basis of their latent period of spore germination. Snow (1949) studied the rate of germination of spoilage fungi at a_w ranges of 1.0 to 0.75. His results are listed in Table 11.4. *Mucor spinosus* did not germinate below an a_w of 0.93. The minimum a_w for spore germination of a species for *Rhizopus* (not listed in Table 11.4) and *Botrytis* is also 0.93.

In Fig. 11.1, cereals are examined microscopically for the presence of molds.

To prevent mold growth on cereals and cereal by-products, moisture content of the food must be low enough. The maximum moisture levels for both short- and long-term storage, listed in Table 11.5, were compiled by Snow *et al.* (1944).

TABLE 11.4.   LATENT PERIOD OF SPORE GERMINATION (DAYS) OF MOLDS AT CONSTANT ERH AT 25°C ON NUTRIENT GELATIN

Fungus	ERH												
	100	95	93	90	88	86	84	82	80	78	76	75	73
Mucor spinosus	1	1	2										
Mucor sp.	1	1	2										
Rhizopus nigricans	1	1	2										
Botrytis cinerea	1	2	2										
Cladosporium herbarum	1	1	2	2	7								
Aspergillus niger	1	1	2	2	3	4	11						
Penicillium wortmanni				2	2	3	4	20					
Aspergillus sydowi				1	2	2	4	7	12	34			
Aspergillus candidus								4	7	9	12	15	

Source: Data taken from Snow (1949).

Courtesy of University of Missouri-Columbia

FIG 11.1.   FOOD MICROBIOLOGIST EXAMINES PETRI PLATE FOR THE
PRESENCE OF MOLDS IN CEREAL GRAINS

DOMINANT SPOILAGE FLORA OF DEHYDRATED FRUITS, VEGETABLES AND CEREAL GRAINS

Mossel and Ingram (1955) listed *Aspergillus, Penicillium, Fusarium, Monilia* and *Rhizopus* as the dominant flora on cereal grains when spoilage occurs during standard conditions. Von Loesecke (1955) found the following

TABLE 11.5.   MOISTURE LEVELS BELOW WHICH MOLD SPOILAGE WILL NOT
OCCUR

Food Product	% Moisture for Short Storage Period	% Moisture for Long Storage Period
Wheat	15.7	14.6
Corn	14.8	13.7
Barley	14.8	13.6
Oats	14.5	13.4
Middlings	14.4	13.1
Bran	14.4	12.8
Peas	14.7	13.3
Soybeans	13.3	11.4

Source: Snow *et al.* (1944).

molds on dehydrated vegetables: *Aspergillus, Mucor, Penicillium, Sporotrichum, Trichoderma, Monilia, Thamnidium, Alternaria,* and *Fusarium.*

MYCOTOXIN PROBLEM

Cereals make up the major source of protein for most of the people of the world. Occasionally, mycotoxins contaminate these important foods. Contamination of corn, for example, may occur in the field and in storage. Mycotoxins produced in cereals are not limited to a particular geographic or climatic region. *Aspergillus, Fusarium, Penicillium* and *Claviceps* are the genera most commonly involved. The mycotoxins occurring include aflatoxins, ochratoxins, penicillic acid, patulin, ergot, zearalenone, citrinin, T-2, tenuazonic acid, kojic acid, and sterigmatocystin. Aflatoxins, patulin, penicillic acid and sterigmatocystin are carcinogens (Hesseltine 1974).

DEHYDRATION PROCEDURES AND THEIR INFLUENCE UPON THE MICROFLORA

Raw Product Quality

Any method of food preservation does not improve upon the quality of raw material; therefore, only high quality raw materials should be used in dehydration of fruits and vegetables. Cereal grain should be dried as rapidly as possible to prevent contamination with mold.

Grading, Selection and Sorting

Grading, selection and sorting of fruits and vegetables will influence the kinds and numbers of microorganisms present on the finished product. Dry sorting of vegetables aids in eliminating spoiled raw materials prior to washing. Dry cleaning eliminates soil from the product (and many microorganisms) prior to washing.

Washing the Product

Potable water should be used in washing the fruit or vegetable prior to processing. In-plant chlorination of 2–5 ppm of residual chlorine will reduce the growth of microorganisms on equipment and on the raw fruit or vegetable.

Peeling

If the raw material needs to be peeled prior to drying, steam or lye at temperatures of 60°C (140°F) may be used. Both operations reduce microorganisms on the product.

Alkali Dipping

Plums (to make prunes) and grapes (to make raisins) may be dipped in alkali to check the skin. This process allows moisture to escape more readily during drying and also reduces flora.

Blanching

Vegetables are blanched at 87.7°–100°C (190°–212°F) prior to dehydration. The temperature of the blanch and the time are dependent upon the maturity of the vegetable. A reduction of 99% of the microflora can occur. Blanching also inactivates enzymes.

Sulfuring

Sulfur dioxide or sulfites are added to many fruits and vegetables to prevent browning of the plant material prior to drying. Sulfites can reduce the numbers of microorganisms. Fruits are exposed to SO_2 by burning sulfur to obtain a level of 1000–3000 ppm depending upon the fruit and the amount absorbed. The sulfur dioxide conserves vitamins C and A.

Sun Drying

Sun drying is limited to a favorable climate and to certain fruits. Raisins, prunes, figs, apricots, pears and peaches can be sun dried. In addition to possible spoilage during the drying process, the product is also open to further contamination from dust, dirt, insects and rodents. All of these would add microorganisms to the dried fruit.

Dehydration

Water is removed from fruits and vegetables using cabinets, tunnels, kilns and freeze dryers. Factors influencing drying in tunnel and cabinet driers are: temperature, relative humidity of the air, velocity of the air flow, and time. Fruits are dehydrated at temperatures of 60°–73.8°C (140°–165°F); vegetables are dried at 57.2°–93.3°C (135°–200°F).

Temperature. Temperatures used in tunnel and cabinet drying range from 57.2° to 93.3°C (135° to 200°F). This range covers the thermal death point of some molds. For example, the thermal death point of *Rhizopus nigricans* is 60°C, for *Monilinia fructicola* it is 52° to 52.5°C, and for *Penicillium digitatum* it is 58° to 58.5°C. Since evaporation is a cooling process, the product does not reach the dry bulb temperature of the drier until the product has a moisture content of about 2 lb water to 1 lb of bone-dry matter.

Sweating or Moisture Adjustment. With tunnel-dried products, some pieces of the dried plant material may have a lower moisture content than others. The dried material is put in a bin so that the overall moisture content comes to a common value. Some microbial growth may occur if the moisture adjustment is not rapid enough.

Pasteurization. Dried fruits such as dates are heated at 65.6°–85°C (160°–185°F) at a relative humidity of 70–100% for 30 to 70 min to destroy pathogens (Fellers 1930).

Packaging. Dehydrated vegetables are packaged in metal containers (5 gal. or No. 10 cans). In-package desiccation keeps the moisture at a very low level and prevents microbial growth.

Storage Time. If the moisture level is kept low in the package, the number of bacteria and molds present in the product die off during storage. The drop in numbers is rapid at first and slower thereafter.

QUALITY OF DEHYDRATED FRUITS AND VEGETABLES

Vegetables

During World War II, vast quantities of vegetables were produced in the United States. The data of Vaughn (1951) are listed in Table 11.6. These samples were taken from 5-gal. lots produced in adapted fruit tunnel driers. The results revealed that dehydrated vegetables can be produced with very low counts, indicating good manufacturing practices.

TABLE 11.6.   VARIATIONS IN TOTAL COUNTS OF COMMERCIAL SAMPLES OF DEHYDRATED VEGETABLES CERTIFIED AS ACCEPTABLE, 1944–45

Product	No. of Samples	Total Count per Gram Min	Total Count per Gram Max	Maximum Finishing Temp °C	Maximum Finishing Temp °F
Onions	53	12,600	7,500,000	60	140
Potatoes	31	<100	22,800,000	68	155
Sweet potatoes	18	<10	47,000	74	165
Cabbage	20	<100	60,000	63	145
Rutabagas	2	40,000	92,000	65	150
Carrots	11	<100	40,000	71	160
Beets	4	60,000	520,000	68	155

Source: Data taken from Vaughn (1951).

Fruits

Because French-packaged dates had been involved in an outbreak of colitis, Fellers (1930) studied dried fruits on American markets. It will be noted by studying data given in Table 11.7 that dates and apples contain the highest microbial counts. Bacteria resembling *Aerobacter aerogenes* (*Enterobacter aerogenes*) were found in only 8 samples. None were contaminated with *Escherichia coli*.

Aspergillus was the most numerous of the molds; *Penicillium* was second and *Mucor* third. Most of the samples did not contain any yeast cells.

Fellers (1930) found that the total numbers of organisms were greatly reduced by pasteurization. On dates the average reduction varied from 93 to 99%. Pasteurizing prunes and seedless raisins yielded similar results. Inconsistent results were obtained in experiments with figs.

Fellers inoculated dates, figs, raisins, and prunes with *Escherichia coli* and with *Salmonella typhi*. A 30-min exposure to 71.1°C (160°F) with a relative humidity of 75% destroyed the *E. coli; S. typhi* was destroyed by 76.7°C (170°F) for 25 min at a high humidity.

Tanaka and Miller (1963) isolated yeasts and molds associated with spoiled dried prunes. They isolated 62 strains of yeasts. The species were as follows:

> 21 strains of *Saccharomyces rouxii*
> 11 strains of *S. mellis*
> 8 strains of *Torulopsis magnoliae*
> 5 strains of *T. stellata*
> 4 strains of *Candida krusei*
> 3 strains of *Trichosporon behrendii*
> 2 strains of *Pichia fermentans*
> 2 strains of *P. membranaefaciens*
> 2 strains of *C. chalmersii*
> 1 strain each of the following:
>   *S. rosei, S. cerevisiae, Sporobolomyces*
>   *roseus,* and *C. parapsilosis.*

Fewer different species of molds were among the 124 strains isolated from the fermented prunes. *Aspergillus glaucus* was the leading species with 56 strains. According to Thom and Raper (1945), *A. glaucus* is commonplace in sweetened products, including jams, jellies, soft sugars, honey, soft candies; in salted products such as meats, pickles; and in dried foods where preservation is a complex of sugars and acid (dried fruits). Tanaka and Miller's data certainly confirm Thom and Raper's statement. They also found 18 strains of *A. niger*, 41 strains of *Penicillium* sp., 4 other strains of *Aspergillus*, 2 strains of *Alternaria* and one strain each of *Monilia* sp., *Mucor* sp. and *Chaetomella* sp.

TABLE 11.7. MICROORGANISMS IN AMERICAN DRIED FRUITS AS PURCHASED

Dried Fruit	Av Moisture Content (%)	Bacteria (per g)	Molds (per g)	Yeast (per g)	No. Samples Containing Lactose Fermenters
Dates, Iraq bulk	16.1	12,300	18,400	20	3
Dates, Iraq packaged	17.1	3,970	35,800	55	2
Dates, Calif. packaged	15.6	760	700	100	0
Dates, Calif. glass-packed	44.0	320	0	0	0
Dates, French Algerian packaged	24.0	18,000	42,000	16,000	1
Figs, Smyrna packaged	9.6	420	20	3	0
Figs, Calif. packaged	15.1	210	200	10	0
Prunes, bulk	15.0	25	15	0	0
Prunes, packaged	16.3	15	6	0	1
Peaches, bulk	13.3	10	0	0	0
Peaches, packaged	15.1	6	0	0	1
Raisins, seedless	15.3	310	340	6	1
Raisins, seeded	17.1	40	90	0	0
Apricots, bulk	11.8	8	17	0	0
Apricots, packaged	13.9	20	0	0	0
Apples, bulk	21.9	40	10	10	0
Cranberries, packaged	6.0	0	3	0	0

Source: Fellers (1930).

Both yeasts and mold isolates studied exhibit osmophilic characteristics. *Saccharomyces rouxii, S. mellis,* and *Torulopsis stellata* grew and fermented media with 70% soluble solids but failed to grow in concentrations of 75% soluble solids. Strains of *Aspergillus glaucus* were more strongly osmophilic than were *A. niger* and *Penicillium sp.*

In a four-month period at 20°C, strains of *A. glaucus* and *S. rouxii* grew on prunes with an a_w of 0.76 but not at an a_w of 0.69. At 30°C, however, only *S. rouxii* grew at an a_w of 0.85.

METHODS FOR THE MICROBIOLOGICAL EXAMINATION OF DRIED FOODS

The American Public Health Association (1958) published the following methods for examination of dehydrated fruits and vegetables and of cereals and cereal products.

Fruits

Fruits are examined using the following methods:
1. Total Count (Orange Serum Agar, pH 7)
2. Yeasts and Molds (Malt Agar or Potato Dextrose Agar)
   If only yeast counts are desired, add 0.25% sodium propionate to the above agars. See American Public Health Association (1958) for composition of media and methods.

Vegetables

Vegetables are examined by the following methods:
1. Total Count (Orange Serum Agar)
   The pH is adjusted to 7.0. Incubate for 3 days at 32°C.
2. Lactic Acid Bacteria
   Confirm colonies with Gram stain and catalase production. Incubate for 3 days at 32°C.
3. Sporeforming Bacteria
   (a) Flat-sour spores (dextrose tryptone agar). Sample is boiled 5 min. Incubate for 48 hr at 55°C.
   (b) Anaerobes producing H_2S (sulfite agar in tubes). Incubate for 48 hr at 55°C.
   (c) Anaerobes not producing H_2S (liver broth). Tubes are made anaerobic by covering with Vaspar or petrolatum.

Cereals and Cereal Products

Cereals and cereal products are examined using the following methods:

1. Total Aerobic Count (Plate Count Agar)
   Incubate at 32°C for 3 days.
2. Total Anaerobic Count (Plate Count Agar)
   Incubate anaerobically at 32°C for 3 days.
3. Thermophilic Counts (55°C)
   (a) Aerobic total counts (plate count agar) 48 hr.
   (b) Anaerobic total count (liver broth) for 48 hr.
4. Thermophilic Aerobic Spore Counts (Dextrose Tryptone Agar)
   Steam sample and agar (120 ml) for 3 min at 100°C (212°F).
   Distribute cereal-agar mixture in 5 Petri plates and incubate for
   48 hr at 55°C.
5. Flat-Sour Types of Spores (Dextrose Tryptone Agar with Bromcresol Purple)
   Add 20 ml of 1–10 cereal dilution to 100 ml of agar. Shake mixture
   for 3 min in boiling water. Then heat 30 min in flowing steam. Add
   to 5 Petri dishes. Colonies (2–3 mm) with an opaque central spot
   surrounded by a yellow halo are counted.
6. Anaerobic Spores Producing H_2S (Sulfide Spoilage) Sulfite
   Agar
   Twenty milliliters of 1–10 cereal dilution are divided (about 3.3
   ml/plate) among 6 tubes of melted sulfite agar. Heat for 30 min
   in flowing steam. Count black colonies in the 6 tubes. Incubate 5
   days at 55°C. This is the number in 10 g of cereal product.
7. Anaerobic Gas-Producing Spores
   Add 20 ml (of the 1–10 dilution) to 6 tubes of liver broth (about
   3.3 ml per tube). Heat for 30 min in flowing steam. Stratify and
   incubate at 55°C for 5 days. Report number of tubes showing
   gas.
8. Yeasts and Molds
   Plate out sample using potato dextrose agar (acidified), malt agar
   or malt-salt agar. Incubate 3 to 5 days at 32°C.

Interpretation of Results

High counts indicate low quality raw materials and/or poor handling
practices during dehydration. High numbers of lactic acid bacteria would
indicate that the vegetable had undergone some fermentation since these
organisms are not present in large numbers on plant tissues.

The significance of sporeformers in both cereal products and dehydrated vegetables is important when the food is canned. A "rope" spore count on a cereal product can be used to determine whether or not ropy bread may develop.

GLOSSARY

Aflatoxin—A mycotoxin produced by *Aspergillus flavus.*

Blanching—Treating vegetables and fruits to inactivate enzymes with hot water or steam.

Dehydration—To preserve food by removal of water under controlled conditions.

Dry bulb temperature—This is the temperature of air as it enters a dehydrator. The wet bulb temperatures are obtained using a thermometer with a wick wet by distilled water. Knowing these two temperatures, one can determine the relative humidity.

Dry cleaning vegetables—To remove soil by placing vegetables in rotating cylinders and shaking devices.

Equilibrium relative humidity—The relative humidity above a substance in a closed container.

Flat-sour—Pertaining to bacteria that produce acid but no gas in the spoilage of canned foods.

Non-ionizing solute—A solute that does not ionize in water. An example is sucrose.

Osmophilic—Loving a high osmotic pressure.

Pasteurization—To heat a food to destroy pathogenic microorganisms.

Standard conditions—Foods stored at the recommended temperature, relative humidity and in the required packaging.

Thermal death point—The temperature that will kill a microorganism in ten minutes.

Water activity—The amount of water available is expressed on a scale of 1.0 to 0.0. Pure water is 1.0 on the a_w scale. An a_w of 0.7 means that most microorganisms cannot grow.

REVIEW QUESTIONS

1. Why can fruits be dried to a higher moisture content than vegetables can without spoiling?
2. How is water activity measured?
3. What is equilibrium relative humidity?
4. How is ERH related to a_w?
5. At what levels of moisture do carrots and milk start to spoil?

6. What a_w is required to give reasonable protection from spoilage?
7. What is meant by alarm water content?
8. Using a_w for the lowest levels at which growth occurs, state the levels for osmophilic yeast, xerophilic fungi, halophilic bacteria, normal bacteria, normal molds and normal yeasts.
9. What genus is dominant in spoiling cereals?
10. What was the lowest ERH at which fungi grew in Snow's study?
11. What are the moisture contents of cereals below which no mold will grow?
12. List the dominant flora that spoil cereal grains according to Mossel and Ingram?
13. Other than odor and flavor, what is the concern about moldy foods?
14. Discuss the effect of procedures used in dehydration upon the microflora of fruits and vegetables.
15. Why are dried fruits pasteurized?
16. Is it possible to produce dried fruits and vegetables with low microbial counts?
17. Name the dominant osmophilic yeasts (2) and filamentous fungi studied by Tanaka and Miller (1963).
18. Discuss some microbiological tests used to evaluate dehydrated fruits, vegetables and cereal products.

REFERENCES

AMERICAN PUBLIC HEALTH ASSOC. 1958. Recommended Methods for the Microbiological Examination of Foods. American Public Health Assoc., New York.

FELLERS, C. R. 1930. Pasteurized dried fruits. Am. J. Public Health *20*, 175–182.

HESSELTINE, C. W. 1974. Natural occurrence of mycotoxin in cereals. Mycopathol. Mycol. Appl. *53*, 141–153.

LEUNG, W. W., BUTRUM, R. R. and CHANG, F. H. 1972. Food Composition Table for Use in East Asia. Food Policy Nutr. Div., FAO, United Nations, Rome.

MAKOWER, B. and DEHORITY, G. L. 1943. Equilibrium moisture curves for different vegetables. Ind. Eng. Chem. *35*, 193–197.

MOSSEL, D. A. A. and INGRAM, M. 1955. The physiology of the microbial spoilage of foods. J. Appl. Bacteriol. *18*, 233–268.

SNOW, D. 1949. The germination of mold spores at controlled humidities. Ann. Appl. Biol. *36*, 1–13.

SNOW, D., CRICHTON, M. H. G. and WRIGHT, N. C. 1944. Mould deterioration of feeding-stuffs in relation to humidity of storage. Part 2. The uptake of feeding-stuffs at different humidities. Ann. Appl. Biol. *31*, 111–116.

TANAKA, H. and MILLER, M. W.   1963.   Microbial spoilage of dried prunes. Hilgardia *34*, 167–190.

THOM, C. and RAPER, K. B.   1945.   A Manual of the Aspergilli.   Williams & Wilkins Co., Baltimore.

TUITE, J. F. and CHRISTENSEN, C. M.   1957.   Grain storage studies.   XXIV. Moisture content of wheat seed in relation to invasion of the seed by species of the *Aspergillus glaucus* group, and effect of invasion upon the germination of the seed.   Phytopathology *47*, 323–327.

VAUGHN, R.   1951.   Variations in "total" counts of commercial samples of de-hydrated vegetables certified as acceptable, 1944–1945.   Food Res. *16*, 429.

VON LOESECKE, H. W.   1955.   Drying and Dehydration of Foods.   Reinhold Publishing Corp., New York.

12

Microbiology of Canned Foods

SPOREFORMING BACTERIA IN SOIL

Thermophilic Facultative Anaerobic Spores

Although we know that bacterial spores from the soil contaminate vegetables, comparatively little is known about the ecology of sporeformers.

Bacillus stearothermophilus is a thermophilic, facultative anaerobe which can grow in a can. Fields and Chen Lee (1975) studied the distribution of thermophilic aerobic and facultative spores in soils from the east, west and midwest of the United States. Most soils contained from 1 to 270 spores. In some soils counts were greater than 5400 spores/g. They found that 8 out of 11 $D_{121°C}$ values were in the same range as $D_{121°C}$ values listed by Stumbo (1965) for low and semiacid foods.

The spores coming into the plant can "seed" equipment and can germinate, grow, and produce more spores if proper sanitation is not maintained. A review of the importance of flat-sour bacteria to the canning industry was written by Fields (1970). This review should be consulted for a discussion of *B. stearothermophilus* and *Bacillus coagulans*. Gibson and Gordon (1974) stated that although spores of *B. stearothermophilus* occur in soil in all climatic zones, *B. coagulans* spores are relatively scarce.

The relationship of soil constituents to spore counts and the heat resistance of *B. stearothermophilus* were studied by Fields *et al.* (1974). They found spore counts of *B. stearothermophilus* were influenced by types of crops and were correlated positively with Mn, Ca, P and pH of soil from Sanborn Field, Columbia, Missouri. The content of Mg, Ca, K, and cation-exchange capacity was correlated with counts of the thermophilic aerobic counts. Spores produced in Sanborn Field were more heat-resistant than spores formed in greenhouse soil.

Thermophilic, Anaerobic Spores

Small numbers of *Clostridium thermosaccharolyticum* have been found in soil (Smith and Hobbs 1974). Fields and Chen Lee (1975) chose 65°C as the temperature of incubation of the liver broth tubes. This made the conditions partially selective for *C. thermosaccharolyticum*. Most soils from the west, midwest and east contained from 1 to 100 spores. However, some samples had spore counts greater than 500 per g. Fewer anaerobes were present in the soil than facultative anaerobes.

Mesophilic, Anaerobic Spores

Data on heat resistance of soil-produced spores of *Clostridium sporogenes, C. butyricum* and *C. pasteurianum* are not available. By determining the type of *C. botulinum* involved in cases of botulism, the geographic distribution of its spores was established. For example, out of 158 type A cases, 142 cases occurred in states west of the Mississippi; 24 states (most of them in the east) have not reported A outbreaks; type B outbreaks occurred mainly in the east and type E predominated in the Great Lakes region (Department of Health, Education and Welfare 1974).

EFFECTS OF THE ENVIRONMENT IN THE CONTAINER ON GROWTH OF SPOILAGE MICROORGANISMS

Vacuum

Partial vacuums are formed in tin cans after heat processing the foods. Partial vacuums as high as 16–20 in. of Hg are not uncommon. The vacuum itself eliminates the growth of strict aerobes if they survived the heat process. In minute quantities of water, nonsporeformers can also enter tin cans during handling and cooling. Facultative types can also grow in the can and cause spoilage. In Fig. 12.1, a scientist is shown determining the vacuum in a can, an important step in determining the type of spoilage present.

According to Hersom and Hulland (1964), *Byssochlamys fulva* can tolerate a low oxygen tension (20 in. of vacuum). *Penicillium striatum,* which, tolerates a vacuum of 12 to 15 in., has been isolated from spoiled canned blueberries.

These filamentous fungi are the exception, not the rule, as growth of most fungi is prevented by a partial vacuum.

FIG. 12.1. FOOD MICROBIOLOGIST TAKES THE VACUUM OF HOME-CANNED GREEN BEANS BEING STUDIED IN THE DEPARTMENT OF FOOD SCIENCE AND NUTRITION AT UNIVERSITY OF MISSOURI-COLUMBIA

Type of Bacterium

The two flat-sour types of bacteria causing spoilage in canned foods are *B. stearothermophilus* and *B. coagulans*. Both produce acid (hence the name sour) and produce no gas (and the can remains flat). Since both are facultative, they grow well in partial vacuums in cans.

Anaerobes produce gas so that the can explodes if the growth of the bacteria continues. Of course, the vacuum assures growth of anaerobes which survive the heat treatment.

pH

The types of bacteria which will grow in canned foods are also influenced by the pH of the food. Not only does this affect growth, but also it affects the rate of kill of spores during the heating process. Foods with pH values less than pH 4.5 can be processed at 100°C because bacteria are more easily killed.

Temperature of Storage

Flat-sour spores, for example, may not be killed during the heat process. If the food is then stored below the growth temperature of the bacteria, they will die. This process is called autosterilization (Hersom and Hulland 1964).

METHODS OF CANNING AND THEIR RELATION TO SPOILAGE OF LOW-ACID VEGETABLES

Dry Sorting

Sorting of vegetables (especially tomatoes) is easier before washing them. Also if much of the soil can be removed from root crops before washing, fewer soil and bacterial spores will accumulate.

Fluming

Many vegetables are conveyed into the cannery and to various parts of the cannery in running water. A supply of potable water should be added so that spores do not build up in these waters.

Washing

The vegetables should be washed with high-pressure, potable water sprays to remove all soil residue. If in-plant chlorination is used, the National Canners Association (1968) recommends 4–7 ppm residual chlorine at the point of application of water to the equipment.

Re-Use of Water

If re-use of water is necessary, only clean water should be used for washing vegetables. This water should be chlorinated prior to applying to the raw product. Fresh, potable water should be used in the final wash.

Conveying of Raw Product or Blanched Product

Conveyor belts should be sprayed with chlorinated water (4–7 ppm) to prevent buildup of spores on the equipment.

Blanchers

Blanchers, because of the heat and the materials extracted from the vegetables, provide an environment favorable for spore formation. If plant residue and heat remain in this piece of equipment overnight, spores may develop which will feed into the product the next day. Typical cells of *B. stearothermophilus* are shown in Fig. 12.2.

Fillers

Again, the product is held hot in the fillers during operation. If there are steam leaks and faulty cleaning, there may be spore formation in the equipment. Flat-sour bacteria can produce spores in the plant if heat and substrate are available.

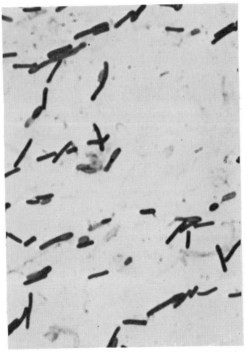

Courtesy of University of Missouri-Columbia

FIG. 12.2. SPORULATING CELLS OF *BACILLUS STEAROTHERMOPHILUS*

B. stearothermophilus can sporulate in blanchers, if the blanchers are not cleaned properly and if they are allowed to remain warm.

Cooling

Minute quantities of water may be drawn into the can after heat processing at the time a vacuum is produced. Nonsporeformers can thus cause spoilage at this point in canning operations. For this reason, cooling water should be chlorinated to 5 ppm residual chlorine at inlet and 0.5 ppm at outlet in cooling canals, tanks and rotary continuous coolers (National Canners Association 1968).

Can Handling

Troy *et al.* (1963) studied the effect of leakage on canned food spoilage. They found that the type of spoilage can be grouped broadly into two categories: namely, factors relating to can making and in-plant processing operations which include contaminated can handling equipment and rough handling.

SANITATION

Potable Water

Potable water is water that is fit to drink. It contains no harmful bacteria or chemicals. Such water should be available throughout the canning plant.

Cleaning

As stated by the Association of Food Industry Sanitarians and National Canners Association (1952), canning plants should be physically, chemically, and bacteriologically clean. To accomplish this, a comprehensive cleanup program with adequate equipment must be in effect. Equipment must be thoroughly cleaned before chlorination. (See the above book for details and also consult the *Laboratory Manual for Food Canners and Processors, Vol. 2* for further instructions.)

Chlorination

Residual chlorine content in in-plant chlorination ranges from 4 to 7 ppm. At cleanup, concentrations of 10–20 ppm can be used. Consult National Canners *Laboratory Manual for Food Canners and Processors, Vol. 2* (1968) for list of foods and the chlorine concentrations which might cause off-flavors.

pH CLASSIFICATION OF FOODS

Food Groups

Cameron and Esty (1940) grouped foods according to their pH. They suggested four groups.

Group 1. *Low-Acid* (pH 5.0 and higher): Meat products, marine products, milk and certain vegetables.

Group 2. *Medium-Acid* (pH 5.0 to 4.5): Meat and vegetable mixtures, specialties such as spaghetti, soups, and sauces.

Group 3. *Acid* (pH 4.5 to 3.7): Tomatoes, pears, figs, pineapple and other fruits.

Group 4. *High-Acid* (pH 3.7 and below): Pickles, grapefruit, citrus juices and rhubarb.

Spoilage Microorganisms Associated with These Food Groups

Bacteria Spoiling Groups 1 and 2

(a) Thermophiles

Flat-sour, *B. stearothermophilus*

Anaerobes not producing H_2S, *C. thermosaccharolyticum*

Anaerobes producing H_2S, *Desulfotomaculum nigrificans*

(b) Mesophiles

Anaerobes

    C. botulinum

    C. sporogenes

    C. butyricum

    C. pasteurianum

Facultative anaerobes

    B. licheniformis

    B. cereus

    B. megaterium

Bacteria Spoiling Group 3

(a) Anaerobes (sporeformers)

    C. pasteurianum

    C. butyricum

(b) Facultative Anaerobes (sporeformers)

    B. coagulans

Microorganisms Spoiling Group 4

(a) Yeasts

(b) Molds (*Byssochlamys fulva, Penicillium striatum*)

The Botulism Problem

The lowest pH at which the spores of *C. botulinum* A and B grow is 4.8. At 4.6, spore germination is inhibited (Schmidt 1964). Types A and B are the most heat-resistant and most frequently the cause of botulism from canned foods.

Bacterial spores are most heat-resistant at pH values above 4.5, becoming the most resistant at pH 7. For this reason, foods with pH values greater than 4.5 are processed under steam pressure at 115.6°–121°C (240°–250°F).

Since type E *C. botulinum* spores are not heat-resistant, they are only a threat if they enter canned food through faulty can seams after thermal processing.

Canned vegetables are the leading cause of botulism in the United States. For the years 1899–1973, vegetables were involved in 17.4% of the cases. Unknown causes were responsible for 66.7% of the outbreaks. During the period of 1950–1973, home-canned green beans accounted for 12 outbreaks and home-canned mushrooms for 10 outbreaks (Department of Health, Education and Welfare 1974).

Between 1950 and 1973, botulism from commercially canned foods occurred in liver paste, tuna fish, potato soup and peppers. Over the years, commercial canning has had a very good record when one considers all the billions and billions of cans involved.

SPORE PROBLEMS IN FOOD INGREDIENTS

The National Canners Association (1968) recommends the spore counts standards for sugar and starch as shown in Table 12.1.

Although no standards have been given for flour, spices, condiments, molasses, cocoa, milk (dried), these should be low in spore numbers; otherwise, these ingredients, like sugar and starch, will contaminate the product and can cause spoilage. Bacterial spores die logarithmically so that the greater the number present the more time it will take to kill all of them.

TABLE 12.1.   STANDARDS FOR SUGAR AND STARCH

| | Spore Counts per 10 g Should Not Exceed | | | |
| | Sugar | | Starch | |
Type of Bacteria	Max	Avg	Max	Avg
Total thermophilic	150	125	150	125
Flat-sour	75	50	75	50
Thermophilic anaerobes not producing H_2S	3/5		3/5	
Sulfide spores	2/5		2/5	

Source: National Canners Association (1968).

HEAT RESISTANCE OF BACTERIAL SPORES

The following influence the heat resistance of bacterial spores:

The Age of the Spore. An aged spore is less heat-resistant. Spores held in a suspension at refrigerator temperature show this response.

The Species. Thermophilic spores of *Bacillus stearothermophilus* are the most heat-resistant. A cross section of *B. stearothermophilus* is shown in Fig. 12.3.) Next in heat resistance are those of *Clostridium thermosaccharolyticum.* (See *D* values described under Rate of Destruction Studies given below.

The Temperature at Which the Spore Was Formed. Spores formed at higher temperatures are more heat-resistant than those developed at a lower temperature by the same species.

Sporulation Media. Anaerobic spores formed on raw meat are not as heat-resistant as spores produced on cooked meat.

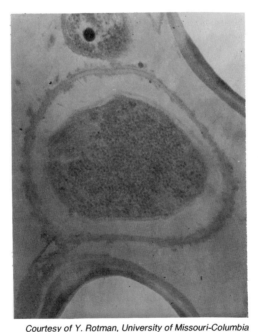

Courtesy of Y. Rotman, University of Missouri-Columbia

FIG. 12.3. CROSS SECTION THROUGH A SPORE OF *BACILLUS STEAROTHERMOPHILUS*
The spores of this bacterium are exceedingly heat resistant. × 125,000.

The Medium in Which Spores Are Heated. Spores in fats are killed by oxidative processes and are killed more slowly than in wet heat. Spores in an acid environment are more easily killed than in an environment near pH 7.

The Recovery Medium. The more nutritious the recovery medium, the more spores can be found after heating.

Dry and Moist Heat. Bacterial spores are more resistant to dry heat than to moist heat.

METHODS OF MEASURING HEAT RESISTANCE OF SPORES

Thermal Death Time Studies

The classical method of studying the heat resistance of bacterial spores is to prepare a spore suspension with a plate count of 10^4–10^5 spores per milliliter. The reader is directed to Fig. 12.4 and 12.5 which illustrate the appearance of slides made before and after cleaning cultures of a spore-former for thermal death time studies. Seal 2 ml of the suspension in pyrex glass tubing (ID of 7 mm with OD of 9 mm). Multiple tubes are placed in small wire baskets which are lowered into an oil bath set at 110°C (230°F). At various time intervals, baskets are removed and cooled rapidly.

The tubes are broken aseptically and the contents added to an appropriate culture medium. This process is repeated at temperatures of 113°, 115.6°, 118.4° and 121°C (235°, 240°, 245° and 250°F). At each temperature there is a time when all the spores are killed and a time when the spores survive. These two times should be as close as possible while maintaining a point of growth and no growth. The data are plotted on semilog paper with time on the log scale and temperature on the arithmetical scale. Points are plotted for all temperatures. The best straight line passing above the growth times and below the kill times is drawn.

F **Value.** If a specific temperature is not mentioned, it is understood that the temperature is 250°F. Otherwise, it may be written as $F_{240°F}$. The *F* value is the time to kill all the spores at a designated temperature. This value is determined from the thermal death time curve.

z **Value.** The slope of the curve in °F is the *z* value.

F **and** *z* **Values.** These values are used in process calculations (time-temperature) for canned foods.

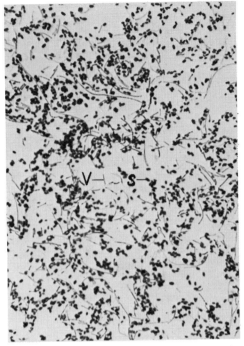

Courtesy of Y. Rotman, University of Missouri-Columbia

FIG.12.4. SPORES (S)   AND   VEGETATIVE
CELLS (V) OF THE ROUGH VARIANT OF *BA-
CILLUS STEAROTHERMOPHILUS* PRIOR TO
CLEANING FOR THERMAL DEATH TIME STUD-
IES

Rate of Destruction Studies

In this case, spores are heated as in the thermal death time studies, but at only one temperature. The suspension is plated with an agar medium. The numbers of survivors are plotted on the log scale and time on the arithmetical scale.

D Value. The time required for the curve to traverse one log cycle is defined as the D value. One D value is 90% kill. Since a D value is not influenced by spore concentration as an F value is, D values are good ways to compare heat resistance.

Stumbo (1965) listed $D_{250°F}$ values for spoilage bacteria which spoil group 1 and 2 foods, as follows:

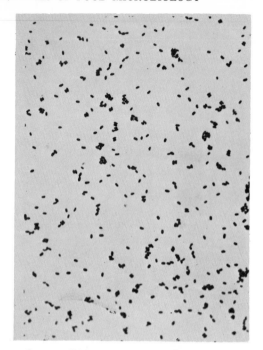

Courtesy of Y. Rotman, University of Missouri-Columbia

FIG. 12.5. SPORES OF THE ROUGH VARIANT OF *BACILLUS STEAROTHERMOPHILUS* AFTER CLEANING FOR THERMAL DEATH STUDIES

Bacterium	D (min)
B. stearothermophilus	4–5
C. thermosaccharolyticum	3–4
D. nigrificans	2–3
C. botulinum A and B	0.1–0.2
C. sporogenes	0.15–1.5
B. coagulans	0.01–0.07

For group 3 foods, the thermal resistance of the spores is lower; therefore, a temperature of 212°F is used. Stumbo (1965) gave D values for the following organisms:

Bacterium	$D_{212°F}$
B. polymyxa and B. macerans	0.10–0.50
C. pasteurianum	0.10–0.50

Again, because of the acidity of group 4 foods, the D values for this group are at 150°F. Yeasts, molds, *Lactobacillus* spp. and *Leuconostoc* spp. have D values of 0.5 to 1.0.

USE OF D, F AND z VALUES IN THE CANNING INDUSTRY

D Values

A reduction of the spore population to 12 D values is required for foods with pH values higher than 4.5 (low-acid and semiacid foods).

F and z Values

These values are used in calculating process times by the General Method and by Ball's formula method of process calculations for canned foods.
These calculations are not in the scope of this book.

SPOILAGE OF CANNED FOODS

As shown in Fig. 12.6, a microbiological examination may be used to determine the causative organisms. The reasons for spoilage follow.

Courtesy of University of Missouri-Columbia

FIG. 12.6.   MICROBIOLOGICAL EXAMINATION OF HOME CANNED FOODS

Underprocessing

If spores are present in large numbers, some survive the heat treatment and the food is called underprocessed. Excessive numbers of spores are caused by the following factors:

1. **Spore Accumulation.** Spore accumulation on plant equipment is due to growth and sporulation. These are facultative spore-formers because the plant environment does not support the growth of anaerobes.
2. **Ingredients.** As discussed before, sugar, starch and spices can be a source of both facultative anaerobic and anaerobic spores.
3. **Poor Washing.** If soil is not removed from the raw materials, the product will have an excessive number of spores.
4. **Combination(s) of the Above.**
5. **Improper Retorting.** Air plus steam can give the correct pressure without the correct temperatures. If thermometers and pressure gauges are not accurate, lower temperature may occur and the operator will not know about it.

Leakage Through Seams

Nonsporeforming bacteria may enter the cans through seams during: (a) cooling, (b) handling.

If pressure-processed cans contain viable cocci or nonsporeforming rods, this is evidence that the product was contaminated after the heat process.

Nonmicrobial Spoilage

The acid of some products may react with the metal in the can to produce hydrogen. This causes a swell. *A swell should be considered dangerous* to health even if caused by hydrogen.

GLOSSARY

Autosterilization—Holding a processed can below the growth temperature of flat-sour bacteria, allowing the viable spores to die with time. For autosterilization to occur, product is underprocessed.
Blanchers—This equipment used to blanch vegetables is usually steam-operated.

Botulism—A food poisoning caused by *Clostridium botulinum* when the vegetative cells produce a very potent exotoxin.

Cation-exchange capacity—The capacity of soil to hold various cations. Soils with more clay have a higher cation-exchange capacity than do sandy soils.

D **value**—The time it takes to kill 90% of the microbial population.

Die logarithmically—A plot of times vs number of survivors (on log scale) will usually be a straight line. This is evidence that bacteria die logarithmically.

Dry sorting—Sorting vegetables before washing. Some rots are more easily detected dry than wet.

F **value**—A value determined on a thermal death time curve. It is the time to kill all the spores at a designated temperature, usually 121.1°C (250°F).

Facultative anaerobe—A bacterium able to grow either aerobically or anaerobically.

Fillers—Machines used to aid the filling of the tin cans.

Flumes—Metal troughs used to transport vegetables by water throughout the cannery.

In-plant chlorination—Chlorination of cannery water at 4–7 ppm residual chlorine to be applied to raw materials and equipment.

Partial vacuum—An incomplete vacuum formed in the tin can is usually in the range from 1 to 20 in. of Hg.

Potable water—Water suitable to drink. It contains no harmful bacteria or chemicals.

Residual chlorine—Free chlorine after all organic matter has been oxidized.

Retort—Same as an autoclave. Equipment used to sterilize canned food.

Sanborn Field—A special experimental field at the University of Missouri, Columbia, Missouri, where crops have been grown since 1888.

Spores—Bodies formed in cells of bacilli and clostridia that resist heat, cold and dryness.

Underprocessing—Heat treating the canned food without sterilization.

z **value**—A value taken from a thermal death time curve. It is the number of degrees Fahrenheit to traverse one log cycle. This value and the *F* value are used in process calculations.

REVIEW QUESTIONS

1. Why study spores in soil?
2. Are thermophilic spores numerous in the soil?
3. Of what significance are thermophilic spores?
4. Are there more thermophilic facultative anaerobic spores or thermophilic anaerobic spores in the soil?
5. Does the treatment of soils (crops and fertilizers) have any influence upon the number of spores in soil?
6. Are all types of spores of *Clostridium botulinum* found in the same geographic location?
7. What types of *C. botulinum* have the most heat-resistant spores?
8. What kinds of bacteria can grow in a partial vacuum?
9. Can any filamentous fungi grow in tin cans?
10. What kinds of bacteria cause flat-sour spoilage? Hard swells in canned foods?
11. List canning operations and state the effect these have on spoilage bacteria.
12. What is in-plant chlorination? What are the recommended levels of chlorine to use in in-plant chlorination?
13. How does leakage spoilage occur?
14. What is potable water?
15. What concentration of chlorine is recommended during cleanup?
16. What is the pH classification of foods? Give the groups and examples of foods in each group.
17. List microorganisms associated with spoilage in each group.
18. Discuss the botulism problem.
19. Give standards for spore counts on sugar and starch.
20. List factors influencing the heat resistance of bacterial spores.
21. Discuss the method of conducting a thermal death time study. How do you plot the results? What factors do you get from a TDTC? Of what use are these factors?
22. Discuss the method of conducting a rate of destruction study. How do you plot the results? What factor do you obtain from this curve?
23. List D_{250} values for the following bacterial spores: *B. stearothermophilus, C. thermosaccharolyticum, D. nigrificans, C. botulinum* A and B, *C. sporogenes,* and *B. coagulans*.
24. To compare the heat resistance of spores of two species of bacteria, should an F or D value be used? Why?
25. Discuss spoilage of canned foods.

REFERENCES

ASSOCIATION OF FOOD INDUSTRY SANITARIANS and NATIONAL CANNERS ASSOCIATION. 1952. Sanitation for the Food-Preservation Industries. McGraw-Hill Book Co., New York.

CAMERON, E. J. and ESTY, R. J. 1940. Comments on the microbiology of spoilage in canned foods. Food Res. *5*, 549–557.

DEPARTMENT OF HEALTH, EDUCATION AND WELFARE. 1974. Botulism in the United States, 1899–1973. Center for Disease Control.

FIELDS, M. L. 1970. The flat-sour bacteria. Adv. Food Res. *18*, 163–217.

FIELDS, M. L. and CHEN LEE, P. P. 1975. Spores of thermophilic bacteria in soil. J. Food Sci. *40*, 384–385.

FIELDS, M. L., CHEN LEE, P. P. and WANG, D. 1974. Relationship of soil constituents to spore counts and heat resistance of *Bacillus stearothermophilus*. Can. J. Microbiol. *20*, 1625–1631.

GIBSON, T. and GORDON, R. E. 1974. Genus *Bacillus. In* Bergey's Manual of Determinative Bacteriology, 8th Edition. R. E. Buchanan and N. E. Gibbons, (Editors). Williams & Wilkins Co., Baltimore.

HERSOM, A. C. and HULLAND, E. D. 1964. Canned Foods: An Introduction to Their Microbiology. Chemical Publishing Co., New York.

NATIONAL CANNERS ASSOCIATION. 1968. Laboratory Manual for Food Canners and Processors, Vol. 1 and 2. AVI Publishing Co., Westport, Conn.

SCHMIDT, C. F. 1964. Spores of *C. botulinum:* formation, resistance, germination. *In* Botulism Proceedings of a Symposium. USDHEW. Cincinnati, Ohio.

SMITH, L. D. S. and HOBBS, G. 1974. Genus *Clostridium. In* Bergey's Manual of Determinative Bacteriology, 8th Edition. R. E. Buchanan and N. E. Gibbons, Editors. Williams & Wilkins Co., Baltimore.

STUMBO, C. R. 1965. Thermobacteriology in Food Processing. Academic Press, New York.

TROY, V. S., BOYD, J. M. and FOLENAZZO, J. F. 1963. Spoilage of Canned Foods Due to Leakage. Continental Can Co., Chicago.

13

Microbiology of Chemically Preserved Foods and Use of Ultraviolet Light in Food Establishments

HIGH SOLIDS-HIGH ACID FOODS

Jelly

A jelly is a semisolid food made from 45 parts of a fruit ingredient and 55 parts by weight of sugar. The fruit ingredient also contains natural sugars to yield a product high in sugar. An optimum concentration of sugar for jelly is 67.5% (Desrosier and Desrosier 1977). The optimum pH is 3.2. From material we discussed previously about osmotic pressure and a_w, we know that the osmotic pressure at 67.5% sugar is up and the a_w down. In fact, the a_w is about 0.96. The combination of a_w 0.96 and pH 3.2 means that most microorganisms are excluded by these factors.

Jams, Fruit Butter and Marmalade

These foods are also high in sugar and low in pH so that the environment of these foods is not conducive to the growth of most microorganisms.

VACUUM

The foods listed above are spoiled by mold growth which is most easily prevented by sealing the products with a vacuum (the way it is done commercially) or by sealing the jars with paraffin (as it is done at home). Most molds (with the exception of those that spoil canned food) cannot grow without atmospheric oxygen.

Sweetened Condensed Milk

This product is high in solids. About 18 lb of sugar are added to 100 lb of milk and the mixture is concentrated to 70% solids. Sweetened condensed milk is not sterilized and is subjected to only a slight heat treatment; the high solids act as the preservative (Desrosier and Desrosier 1977). A vacuum is produced in the can prior to sealing. According to Frazier (1967), spoilage may occur from growth of *Torulopsis lactis-condensi* (with gas formation from the sucrose present) and of *T. globula. Aspergillus repens* or *Penicillium* causes buttons (growth) on the surface of the milk.

FOOD ADDITIVES

Spices

Spices are added to foods to give them flavor. In olden times, spices were thought to have preservative effects. Today, we know that some spices contain chemical compounds which are germicidal. These compounds are listed below.

1. **Cinnamic Aldehyde.**

This compound is found in cinnamon (Frazier 1967).
2. **Eugenol.** $CH_2=CHCH_2C_6H_5(OCH_3)OH$
   Eugenol is present in cloves (Frazier 1967).
3. **Allyl Isothiocyanate.** $CH_2=CHCH_2NCS$
   This bactericidal compound is a constituent of the volatile oil in mustard (Stecher *et al.* 1968).
4. **Acrolein.**

This compound was isolated from garlic and onions by Frazier (1967).
5. **Butylthiocyanate.** $C_4H_9N=CS$
   Frazier (1967) found this compound in horseradish.

Spices contain their own microflora. Yesair and Williams (1942) examined untreated spices. Their data are listed in Table 13.1.

Water extracts of various spices do not effectively inhibit the growth of microorganisms. However, essential oils (steam distillable) obtained from ground mustard, cloves, and cinnamon display a fair degree of bactericidal activity. The relative activity of some oils is as follows:

Volatile oil of mustard has the strongest preserving power. Oil of cloves, oil of thyme and oil of bay leaves follow in descending order of microstatic activity (Weiser *et al.* 1971). Since spice oils are more concentrated than ground spices, they are more effective against the microflora. They are selective in their action. Weiser *et al.* (1971) also found that molds are the most sensitive, yeasts are intermediate in sensitivity and bacteria, especially the sporeformers, are the least sensitive to the germicidal effects of the spice oil.

Acids

Acids have many uses in the food industry. The following is a partial list. Some of these are discussed more fully under Preservatives, which follows.

Benzoic acid, in the form of the acid or the sodium salt, is a preservative of acid foods.

Sulfur dioxide (SO_2) is used on foods to be dehydrated. In contact with water, sulfurous acid is formed.

Carbonic acid (CO_2 in water) is used in beverages and is maintained under pressure.

Propionic acid, in the form of calcium and sodium salts, delays mold growth.

Sorbic acid is a fungistat in apple juice, bread and/or cheese to prevent the growth of molds.

Acetic acid is a component of vinegar which is used in pickling foods. Acetic acid is more toxic to molds, bacteria and yeasts than either lactic acid

TABLE 13.1. MICROBIAL CONTENT OF SPICES

Kind of Spice	37°C Counts per Gram
Whole allspice	106,000–1,000,000
Ground allspice	64,000–1,950,000
Whole cloves	190–12,000
Ground cinnamon	12,500–36,000
Crushed cinnamon	8,000–90,000
Ground red pepper	2,190,000–12,500,000
Ground black pepper	1,200,000–16,300,000
Ground garlic powder	90,000

Source: Yesair and Williams (1942).

or HCl at the same pH. Acetic acid is more inhibitory to selected microorganisms in direct proportion to the amount of acid used.

Commercial strength vinegar is usually 5% acetic acid. A lN CH_3COOH solution is approximately 6%, which results in a pH of 2.36.

If 0.5% of acetic acid is formed in an alcoholic fermentation before the fermentation has proceeded very far, the growth of *Saccharomyces cerevisiae* will be inhibited (Cruess 1958). Vinegar bacteria are usually inhibited by the addition of 125 ppm SO_2 before fermentation of fruits by *S. cerevisiae*.

Levine and Fellers (1940) found that acetic acid inhibited the growth of bacteria at pH 4.9, *S. cerevisiae* at pH 3.9, and *A. niger* at 4.1. They determined that the titratable acidities corresponding to the above pHs were 0.04, 0.59, and 0.27%, respectively. Furthermore, the germicidal pHs of acetic, citric, HCl, lactic, malic and tartaric acids were found to be 4.37, 3.87, 3.80, 2.43, 3.74, and 3.65, respectively.

Apparently, the action of the highly dissociated mineral acid is due to the $[H^+]$ whereas the germicidal properties of organic acids are due to the pH plus the nonionized molecule which can penetrate the cell.

Preservatives

1. Benzoic Acid. This compound or its salts are added to acid foods to prevent the growth of fungi such as *S. cerevisiae*. A concentration of 0.1% is permitted as a preservative in acid foods like pickles or apple juice. These compounds are more effective when the pH of the food is 4.0 or below (Nickerson and Sinskey 1972). *S. cerevisiae* fails to grow in 0.3% and is completely destroyed in a 3% solution.

The mode of action of benzoic acid is as follows:

(a) It is competitive with the coenzyme for the enzyme protein. In the case of glucose or lactic dehydrogenase, the inhibition of these enzymes can be alleviated by the addition of the coenzyme.

(b) Nicotinic acid interferes with the inhibition by benzoate since the chemical structures of the chemical compounds are similar in shape.

(c) High concentrations appear to interfere with the functioning of the membrane of the cell.

In the human body, benzoic acid is eliminated through the urine after combination with glycine to form hippuric acid and glucuronic acid.

According to Frazier (1967), benzoic acid and benzoates have been only moderately successful as preservatives for fish. This probably is linked to the pH of the fish. Benzoic acid is effective at acid pH values, becoming less effective as the pH nears 7.

2.   Sulfurous Acid. The effectiveness of SO_2 is thought to be associated with the formation of sulfurous acid, thus lowering the pH. For SO_2 to be effective, water must be present.

It is also suggested that the SO_2 reduces the S-S linkages in enzymes. At an acid pH (which permits yeast growth), SO_2 or sulfurous acid can penetrate the cell of a yeast and can disrupt normal alcoholic fermentation. The added sulfite reacts with acetaldehyde to form a compound which the enzyme that makes alcohol cannot attack.

Sulfurous acid, SO_2, or sulfites of sodium, potassium or calcium can be used as food preservatives. Sulfurous acid inhibits the growth of yeasts, molds and aerobic bacteria. The SO_2 is lost during storage of foods. Foods in which SO_2 may be found are dehydrated beans, cabbage, carrots, sweet corn and potatoes.

Sulfites may combine with aldehydes and ketones and simple sugars containing these groups. In wines and dried fruits, sulfites become fixed. The toxicity of sulfites with aldehydes and ketones is no greater than that of free sulfurous acid.

Upon ingestion by man, sulfurous acid and sulfites are rapidly oxidized and eliminated in the urine as sulfates.

Because sulfites conceal incipient putrefaction by removing the odor of decay in meat and by keeping the meat a bright red color, their use is not allowed in meat products such as hamburger. Since sulfurous acid and sulfite destroy thiamin, they should not be used in foods with high thiamin content.

3.   Propionic Acid. Calcium or sodium propionate is an effective chemical compound used against molds and many undesirable bacteria found in food products.

It may be used on butter wrappings, on butter tubs, on smoked fish fillets, and on smoked meats. A concentration of $\frac{1}{10}$ of 1% protects fruit fillings from mold growth for long intervals due to the pH of 3.5.

Since propionic acid is a normal constituent in the body, resulting from the metabolism of fatty acids, it is not toxic. The mode of action is due to the competition with certain amino acids for space on active enzyme groups (for example, in the synthesis of pantothenic acid).

4.   Sorbic Acid. Sorbic acid may be incorporated into wrapping material to prevent mold growth, especially on cheese. Sorbic acid has a low toxicity factor. It is metabolized to water and CO_2. It is most effective against yeast at an acid pH. Sorbic acid is used at 0.1% in the same manner as are the benzoates. This compound is effective against film yeasts and may be used in preventing such growth on the surface of fermenting cucumbers.

The mode of action of sorbic acid is its interference with dehydrogenase enzyme systems.

5. Salts.

a.   Nitrates and Nitrites. Nitrites are comparable to sodium chloride in germicidal action (inactive certain enzymes). Nitrates alter the oxidation-reduction potential favoring the growth of aerobic bacteria and restricting the growth of anaerobes (Frazier 1967). Both nitrates and nitrites may be added to meats in curing to fix the red color. In an acid environment, these compounds may have a mild bacteriostatic effect on anaerobes (Weiser *et al.* 1971). The USDA has permitted 200 ppm nitrites in cured meats, but this is to be lowered to 150 ppm.

b.   Sodium Chloride. Salt in solution exerts a certain degree of osmotic pressure and in this way, the growth of microorganisms is influenced. Osmotic pressure or a_w depends upon the number and size of molecules in solution. Compounds such as sucrose with larger molecules than salt have a lower osmotic pressure or a_w than salt (NaCl) at the same molar concentration.

The terms isotonic (0.85% sodium), hypotonic and hypertonic refer to the amount of salt in solution with 0.85% being the reference point. Normal saline (0.85%) can be used as a diluent when making platings of bacteria.

Yeasts, as a rule, are very susceptible to salt with the exception of certain strains of *Torula*. Molds, as a rule, are inhibited in salt concentrations of 20–25%. Many bacilli are not able to grow in salt concentrations over 10%. In 1971, Weiser *et al.* noted that staphylococci were inhibited in a 15% solution and killed at 20%. A concentration of 7.5% sodium chloride is used to make media selective for staphylococci. Pathogenic bacteria are more sensitive to salt than are saprophytes.

Clostridium botulinum is inhibited under certain conditions in 10% salt (Weiser *et al.* 1971). A concentration of 0.5–1% increases the thermal resistance of *C. botulinum* spores, but this resistance is lost at 2%.

Many species of aerobes, facultative anaerobes and true anaerobes are able to grow in high concentrations of brines containing large pieces of animal tissues. In 15–20% salt solutions, the growth of salt-tolerant microorganisms is inhibited, even if whole blood is added. Jensen (1942) observed that the growth of these organisms is confined to the interfaces of brine and tissues.

There are several ways in which sodium chloride can influence microorganisms. The sodium ion can combine with protoplasmic anions of the cell and thus exert a toxic effect. The chloride ion may combine with cell protoplasm and cause death of the cell. Salt can interfere with enzyme systems. A high concentration of salts precipitates proteins or so alters the colloids as to stop the biological activity of the proteins. Concentrated salt solutions "salt out" or cause precipitation of bacterial enzymes involved in respiration (probably dehydrogenating enzymes).

Micrococcus, Leuconostoc, Lactobacillus, Pseudomonas, Flavobacterium, Sarcina and *Halobacterium* are genera which are salt-tolerant.

6. **Sugars.** Sugars can alter the growth of microorganisms in foods by plasmolysis. The kind of sugar and its concentration will determine what microorganisms will grow. Concentrations of from 1 to 10% sucrose will inhibit some microorganisms. Most bacteria and yeasts will not grow in 50% sucrose; concentrations of 65–70% will prevent the growth of molds. Sugar is not a sterile product; it contains bacterial spores, for example.

Because of the size of the molecule, different sugars have different bacteriostatic effects at the same concentration (W/V). For example, dextrose at 35–45% is just as bacteriostatic against staphylococci as a 50–60% solution of sucrose. Dextrose and fructose are more effective than either sucrose or lactose. Because more molecules are present in the smaller weight sugars, the osmotic pressure is greater; hence, they are more bacteriostatic.

ULTRAVIOLET LIGHT (UV)

The UV lethal rays (2000–2800 angstroms) are most germicidal at 2650–2660 Å. UV can be used to sterilize or sanitize surfaces, yet it has very low penetrating power. Ordinary window glass filters out some of the light.

UV can destroy spores of thermophiles in sugar if the sugar is spread thinly and is kept in motion during the exposure. UV is used to keep the air free of microorganisms in bakeries because it kills mold spores and spores of *Bacillus mesentericus (B. subtilis)* (which causes ropy bread). Bacterial spores, however, require 5 to 10 times more exposure than do vegetative cells. Pigmented mold spores are more resistant to UV than are nonpigmented mold spores.

UV may produce some undesirable effects in food. For example, it causes fats to undergo oxidation.

UV can be used to sanitize glassware only if the glasses are clean and are exposed long enough.

GLOSSARY

Bacteriostatic—Merely restricting bacterial growth as contrasted to bactericidal which means to kill bacteria.
Buttons—Fungal growth on the surface of sweetened condensed milk.
Condensed milk—Milk with part of the water removed.

Essential oil—A volatile oil that occurs in spices; for example, oil of mustard.

Plasmolysis—The shrinkage or contraction of the protoplasm in a cell caused by the loss of water through osmosis.

Sanitize—To make sanitary.

REVIEW QUESTIONS

1. What factors preserve jellies?
2. What can spoil jams, jellies, and sweetened condensed milk?
3. List the ingredients in spices which are germicidal. From what spices do they come?
4. Do spices have a flora?
5. What acids are used to preserve foods?
6. Give the mode of action of the following as preservatives in foods: salt, sugar, spices, acids.
7. In what foods should sulfites not be used? Why?
8. Propionic and sorbic acids are used to protect what foods? What microorganisms do they inhibit?
9. Discuss salt as a chemical for preserving food.
10. Of what use is UV in the food industry?

REFERENCES

CRUESS, W. V. 1958. Commercial Fruit and Vegetable Products. McGraw-Hill Book Co., New York.

DESROSIER, N. W., and DESROSIER, J. N. 1977. The Technology of Food Preservation, 4th Edition. AVI Publishing Co., Westport, Conn.

FRAZIER, W. C. 1967. Food Microbiology. McGraw-Hill Book Co., New York.

JENSEN, L. B. 1942. Microbiology of Meats. Garrard Press, Champaign, Illinois.

LEVINE, A. S. and FELLERS, C. R. 1940. Action of acetic acid on food spoilage microorganisms. J. Bacteriol. *39,* 499–514.

NICKERSON, J. T. and SINSKEY, A. J. 1972. Microbiology of Foods and Food Processing. American Elsevier Publishing Co., New York.

STECHER, P. G., WINDHOLZ, M. and LEALEY, D. L. (Editors). 1968. The Merck Index, 8th Edition. Merck and Co., Rahway, N.J.

WEISER, H. H., MOUNTNEY, G. J. and GOULD, W. A.   1971.   Practical Food Microbiology and Technology, 2nd Edition.   AVI Publishing Co., Westport, Conn.

YESAIR, J. and WILLIAMS, D. B.   1942.   Spice contamination and its control. Food Res. 7, 118–126.

14

Microbiology of Fermented Foods

FERMENTATION DEFINED

Fermentation is defined here as processes in which biochemical changes are brought about in organic substrates (carbohydrates, proteins and fats) through the action of biochemical catalysts, enzymes which are produced by specific types of microorganisms.

LACTIC ACID TYPE OF FERMENTATION (KRAUT, PICKLES AND OLIVES)

Comparison of Substrates

Production of kraut is discussed by Prescott and Dunn (1949), Desrosier and Desrosier (1977), Cruess (1958), Vaughn (1954B) and Frazier (1967).

The substrates for the fermentation vary somewhat in composition (Table 14.1). Olives have a much higher carbohydrate content than do either cucumbers or cabbages. It is the soluble carbohydrate fractions which are leached out of the plant material and which are converted into lactic acid and other microbial metabolites. Even if the crude fiber content (not involved in the fermentation) is subtracted from the 16.2% (16.2−3.4%), there is still 12.8% carbohydrate left to undergo possible fermentation. Since the percentage of acid in the olive fermentation is not more than the kraut and salt stock cucumbers, much of the carbohydrate is not involved in the fermentation. Similarities of the fermentations may be seen in the percentage of acidity, pH and salt used in the production of these foods as shown in Table 14.2.

TABLE 14.1. COMPOSITION OF OLIVES, PICKLES AND CABBAGE

Food	Carbo-hydrate	Protein	Fat	Moisture
Olives	16.2	1.2	1.2	80.2
Cucumbers	3.4	0.9	0.1	95.1
Cabbage	5.4	1.3	0.2	92.4

Source: Watt and Merrill (1963); Leung (1972).

TABLE 14.2. COMPARISON OF % ACID, pH AND % SALT USED IN FERMENTATION OF OLIVES, KRAUT AND PICKLES (SALT STOCK)

Food	Acid as Lactic (%)	pH	Salt Used (%)
Olives[1]	0.7–1.0[2]	3.8 or less	3–8[3]
Kraut[4]	1.5–2.0	3.4–3.6	2.5
Salt stock[4] cucumbers	0.6–0.8	3.5–3.8	10.0[5]

[1] Source: Vaughn (1954B).
[2] 2–5% sugar (corn) added to brines.
[3] Varies with type of olive; Spanish type 5–6%.
[4] Source: Cruess (1958).
[5] 10% to start and 15% after the fermentation is finished.

Role of Salt

Brines are prepared in the fermentation of kraut, cucumbers and olives. Lactic acid bacteria are salt-tolerant and are present naturally on plant material. Salt restricts the growth of undesirable microorganisms while allowing the desirable species to grow and produce lactic acid.

Levels of Acid Production

The amount of carbohydrates leached from cucumbers is less than that of cabbage for two reasons. First, there are more carbohydrates in cabbage, and second, the cabbage is shredded whereas the cucumbers are not. As shown in Table 14.2, more lactic acid is produced in kraut. About the same amount of acid is produced in olives and cucumbers (salt stock).

There is a special problem with olives, however. Olives are treated with lye to destroy a bitter glucoside. As much as 65% of the total carbohydrate may be lost in this treatment. Corn sugar (glucose) or beet or cane sugar (sucrose) is added so that the total acidity will reach the level listed in Table 14.2.

The pH of these foods is checked during fermentation. An example of apparatus used is shown in Fig. 14.1.

Courtesy of University of Missouri-Columbia

FIG. 14.1. A FOOD MICROBIOLOGIST MEA-
SURES THE ACIDITY OF SAUERKRAUT

The Succession of Microorganisms

Vaughn (1954B) found that the kraut, salt stock, cucumbers and olives were fermented by the same group of microorganisms. He divided the fermentation into three stages as follows:

	Predominating Species
Primary	*Enterobacter aerogenes*
	Enterobacter cloacae
	Bacillus pumilus
	Bacillus megaterium
	Bacillus polymyxa
	Bacillus macerans
Intermediate	*Leuconostoc mesenteroides*
	Lactobacillus plantarum
	Lactobacillus brevis
	Lactobacillus fermenti

Final   *Lactobacillus plantarum*
                           Lactobacillus brevis
                           Lactobacillus fermenti

Defects and Spoilage (Vaughn 1954B)

1. Kraut
(a) Darkening of the kraut due to iron-tannin (chemical) reactions or to the growth of microorganisms.
(b) Pink kraut caused by the growth of *Rhodotorula* yeasts.
(c) Soft kraut resulting from abnormal fermentations due to air pockets and high temperatures.
(d) Rotting of kraut caused by bacteria and molds.
(e) Off-flavors due to an abnormal sequence of the types of bacteria in the fermentation.

2. Salt Stock Cucumbers
(a) Growth of film yeast with a rise in pH. *Bacillus* sp. and *Penicillium* sp. then cause softening of cucumbers.
(b) "Floater" spoilage by *Enterobacter* sp. Elimination of the largest cucumbers helps to control this spoilage.
(c) Ropy brine. *Leuconostoc mesenteroides* produces a polysaccharide. Other microorganisms can contribute to the problem.

3. Olives
(a) "Floaters" caused by *Enterobacter aerogenes* and *E. cloacae.*
(b) Softening due to growth of *Bacillus polymyxa-macerans* group and species of *Penicillium.*
(c) Malodorous fermentation caused by *Clostridium butyricum.* Careful handling of olives is needed until the pH gets below 3.8–4.0 when this type of spoilage does not occur.

LACTIC ACID TYPE FERMENTATION (SAUSAGES)

As discussed in the chapter on meat, the lactics are normally found in ground meats. With salt and spices which are added to the meat in the preparation of summer sausages, these bacteria dominate the fermentation and produce lactic acid which aids in the preservation of this product.

ALCOHOLIC FERMENTATION: BEER (PRESCOTT AND DUNN 1949)

Brewing Defined

Brewing is the production of malt beverages. Beer, ale, porter and stout are examples of these beverages. A malt beverage is prepared from infused grains which have undergone sprouting (malting) and by the fermentation of the sugary solution (wort) by yeasts. A portion of the carbohydrate is changed into ethyl alcohol and carbon dioxide. A normal beer contains dextrin, maltose, glucose, peptones, amino acids, resins, essential oils and tannic acid.

Malt

Malt is prepared by steeping barley and allowing the grain to germinate and grow until the small plant is about 3/4 the length of the grain. The grain is then dried so that the maximum enzyme activity is retained.

Malt Adjuncts

The malt contains some carbohydrate, but additional carbon sources are added in the form of one or more of the following: corn, corn products, corn flakes, corn sugar, rice, sugar and syrups. The addition of carbohydrate reduces the percentage of nitrogen.

Mashing

Mashing is a process which digests the starch to fermentable sugars. This may be done by two procedures:

1. Infusion Method. The temperature is raised from 38° to 50°C for 1 hr. This favors the proteolytic enzymes of the malt. The temperature is then increased to 65°–70°C for starch digestion. Malt contains both an alpha- and beta-amylase. After maximum sugar formation, the temperature is raised to 75°C to destroy the enzymes.

2. Downward Process. The same temperatures of 65°–70°C are used and the procedure repeated lowering the temperature from 70° to 50°C as above, and then inactivating the enzymes. The time sequence is the same in both processes.

3. Important Considerations in Mashing:
   (a) The malt should be ground, but not too fine.
   (b) The starch should be gelatinized.
   (c) The water should contain 200–300 ppm calcium.

(d) The pH should be 5.0–5.2.
(e) Temperature of 60°C produces nitrogen compounds of high molecular weight. A high temperature favors alpha-amylase activity more than beta-amylase.
(f) Of course, the time used for enzyme hydrolysis of substrates as well as the concentration of enzyme and substrate are important.

Yeasts

Yeasts are classified by industry as top or bottom yeasts. Both are *Saccharomyces cerevisiae.*

Fermentation Process

If a bottom yeast is used, the fermentation is complete in 8–10 days at 6°–12°C. If a top yeast is used, the fermentation is complete in 5–7 days at 14°–23°C.

Maturing

The fermented beer is aged two weeks to several months at 0°C. During this time, unstable proteins and resins precipitate. Esters are formed, giving the beer its mellow flavor.

Finishing

The beer is filtered to give a clear liquid which is then cooled and carbonated (0.45–0.52% CO_2) in bottles or cans. Beer may be pasteurized for 20 min at 60°–61°C to kill bacteria which could cause spoilage.

Defects of Beer

Both microbial and nonmicrobial defects occur in beer.

1. Turbidities. These are caused by unstable proteins, starch due to improper hydrolysis, and microorganisms left in the beer due either to lack of clarification or to secondary infections.

2. Infections. *Acetobacter turbidans* (*A. pasteurianus*) causes sourness and turbidity. Sourness may be caused by *Lactobacillus*. Other infections include sarcina sickness. Sourness and turbidity may be caused by *Streptococcus*.

Types of Beer

Lager Beer. Lager is German for stored beer. It is produced by bottom fermentation.

Bock Beer. This heavy, dark-colored beer is high in alcohol.

ALCOHOLIC FERMENTATION: WINE (PRESCOTT AND DUNN 1949)

Raw Materials

Wine is made from grapes. The sugar content of the grapes should be 20–22%. The grapes may be crushed and fermented or the juice may be produced and then fermented.

Treatment of Grapes

If the grapes are crushed but the juice is not produced, either sulfur dioxide (125 ppm) or a sulfite solution is added to prevent the growth of molds and undesirable bacteria like *Acetobacter*. If small volumes of wine are being prepared for home use, juice may be pasteurized and then inoculated with a pure culture of yeasts.

Fermentation

Saccharomyces cerevisiae is the yeast used to produce wine. Although this organism occurs naturally on fruits, a surer way to produce a successful wine is to inoculate the grapes with a pure culture. One can follow the rate of fermentation by measuring the alcohol content and/or the sugar content. As a rule of thumb, about 50% of the sugar content is converted into alcohol until the level of alcohol reaches about 17%. At this concentration, the yeast is killed.

Filtration

The fermented grape material is filtered and the juice clarified.

Bottling and Aging

The wine is bottled and held to age for the production of esters.

ALCOHOLIC FERMENTATION: DISTILLED BEVERAGES

Rum (Prescott and Dunn 1949)

Raw Materials. Run is an alcoholic distillate from the fermented juice of sugar cane, sugar cane syrup, sugar cane molasses or other sugar cane by-products. The product should have the characteristic taste and aroma generally attributed to rum. The alcoholic content is less than 190 proof. Proof divided by 2 equals percentage.

Yeasts. Strains of *Saccharomyces cerevisiae* or *Schizosaccharomyces pombe* are used in the fermentation.

Fermentation. Phosphorus and a nitrogen source are added to blackstrap molasses (12–14% fermentable sugar) or a similar carbohydrate material. The pH is adjusted to 4.0–4.7 with H_2SO_4. A low pH favors a rapid fermentation and produces a light-flavored rum. A lower pH causes a heavy-bodied rum. Strict sanitation must be maintained for the production of a light-flavored rum.

A temperature of 21.1°–35.5°C (70°–96°F) is adequate. The fermentation is completed in 2 days, but is usually run 3 to 7 days.

Distillation. To produce beverages high in alcohol, the mash must be distilled, an important process which concentrates the alcohol.

Mashes which are distilled at high proof produce light-flavored rums. These rums are to be consumed quickly.

Aging. Rums may be aged quickly by passing the rum over charred white oak chips. Normal aging occurs in charred white oak barrels. The flavor, color, and aroma develop during aging due to an increase in esters, organic acids, and solids.

Whiskey (Prescott and Dunn 1949)

Raw Materials. Whiskey is an alcoholic distillate made from fermented mash of grain. Rye whiskey may be made from rye and barley malt. At least 51% of the mash must be rye. For bourbon (corn), the mash must contain more than 51% corn. For example, a typical mash could contain 70% corn, 15% rye and 15% malt.

Yeast. *Saccharomyces cerevisiae* is the yeast which is usually used.

Fermentation Process.

(a) *Sweet Mash Method.* The mash is inoculated directly with the yeast. Less time is involved and higher alcohol yields result.

(b) *Sour Mash Method.* Spent yeasts from tanks previously fermented are mixed with the mash before the new batch is fermented. Also, the manufacturer may grow *Lactobacillus* sp. in the presence of yeasts to inhibit undesirable microorganisms and to contribute to the aroma, flavor, and characteristics of the whiskey.

Distillation. The mash is distilled to obtain higher alcohol levels.

Aging. Whiskey is aged in white oak barrels. This process increases the solids, esters, acids, fusel oil, aldehydes, furfural and color. The solids are extracted from the wood of the barrels.

Brandy (Prescott and Dunn 1949)

Raw Materials. Brandy is a distillate obtained solely from the fermented juice or mash of a fruit so that a wide variety of brandies (apple, peach, cherry, apricot, orange, raisin, etc.) is available.

Yeast. Strains of *Saccharomyces cerevisiae* are used.

Fermentation Process. The fermentation should be kept anaerobic to prevent the growth of aerobic yeasts and spoilage bacteria.

Distillation. The mash or juice is distilled in a pot or continuous still. Brandies usually contain 40–50% alcohol by volume.

Aging. Flavor develops while the brandy is held in white oak barrels.

Miscellaneous (Prescott and Dunn 1949)

Gins. Gins are distilled from a mash or redistilled over juniper berries and other aromatic compounds.

Sloe Gin. This product is a cordial or liqueur with the main characteristic flavoring derived from the sloe berry (wild plum).

Cordials and Liqueurs. These are products obtained by mixing or redistilling neutral spirits, brandy, gin, or other distilled spirits with or over fruit, flowers, plants or pure juices.

VINEGAR (CRUESS 1958; PRESCOTT AND DUNN 1949; VAUGHN 1954A)

Raw Materials

Apples, grapes, oranges or any material which can be used to produce alcohol can also be used to obtain vinegar. If starchy tubers are used, these are crushed, heated to gelatinize the starch, and treated with malt to produce fermentable sugars. They may also be hydrolyzed by heating with H_2SO_4 (0.5–1.0%).

Fermentation No. 1

The first fermentation is anaerobic with *Saccharomyces cerevisiae* used to produce alcohol. In this fermentation if any wild *Acetobacter* is present and produces more than 0.5% acetic acid, the growth of *S. cerevisiae* will be inhibited. For this reason, 125 ppm SO_2 or an equivalent amount of bisulfite is added to the fruit to prevent the growth of mold, wild yeast, *Acetobacter* and lactic acid bacteria (Cruess 1958). Before inoculation, the treated fruit is allowed to stand 2 hr to kill the unwanted microorganisms named above. Normal fermentation ceases at 35°–40°C (95°–105°F). Heat is produced during the fermentation of sugar to alcohol. The conversion of 1 g of sugar per 100 cc raises the temperature 1.2°C (2.16°F). Grape juice with 22% sugar could raise the temperature 26.4°C (47.5°F) with no heat loss (Cruess 1958).

Strict sanitation must be observed. The tanks should be cleaned well and the press and press cloths sanitized.

Fermentation No. 2

This fermentation is strictly aerobic (Vaughn 1954A). Some species of *Acetobacter* are used. Buchanan and Gibbons (1974) list only three species of *Acetobacter*. These are *A. aceti, A. pasteurianus* and *A. peroxydans*.

Orleans Process. Wine (1/4 of the barrel) is added to fresh vinegar in a barrel. The fresh vinegar, in addition to containing the *Acetobacter,* prevents the growth of film yeasts which metabolize the alcohol. The barrel is kept about 3/4 full at 21.1°–26.7°C (70°–80°F). Pasteur recommended that a wooden rack be placed in the barrel so that the *Acetobacter* film could be supported at the surface of the liquid where there was oxygen available for the bacteria to convert the alcohol to acetic acid. By the end of 3 months, the added alcohol is converted into vinegar. At this time, 1/4 to 1/3 of the liquid is withdrawn and new alcohol is added. The process is repeated.

Quick Process. In this process, a tank 10–14 ft high and 48–60 ft in diameter is filled with beechwood shavings (or some other similar material) to create a large surface area. Alcohol at 10% is allowed to flow down through the generator, coming into contact with a film of *Acetobacter* on the surface of the wood. On one pass through the generator, the alcoholic liquid is acidified to 3–3½% acidity. This procedure can be repeated to get the desired acidity.

Submerged Process. In this method, vinegar bacteria are suspended throughout the liquid by blowing air into the liquid. Yields up to 98% are possible.

Aging. Vinegar is aged long enough to obtain ester formation.

Clarification. The vinegar is filtered to obtain a clear liquid.

Pasteurization. Vinegar is pasteurized at 60°C (140°F) for a few seconds to kill *Acetobacter* which produces mother of vinegar and metabolizes the acetic acid.

Defects of Vinegar. In the first fermentation, *Hansenula apiculata* forms enough acetic acid to prevent the growth of *S. cerevisiae*. Film yeast can grow on vats of alcohol before it can be used to produce vinegar. One percent acetic acid added to the alcohol prevents this growth of the film yeast.

PRODUCTION OF FOOD YEASTS

From Waste Plant Materials

Candida utilis and *Candida tropicalis* are grown for human food. Both can metabolize pentoses so that acid-hydrolyzed wood, straw and other waste raw materials containing these sugars can be used as substrates. Both yeasts are high in protein and contain the B complex vitamins.

Whey

Kluyveromyces fragilis can use lactose. Lactose and whey are by-products of cheese manufacture. These by-products have been used to produce yeast protein.

GLOSSARY

Bourbon—A whiskey distilled from a fermented mash containing not less than 51% corn.

Brandy—An alcoholic beverage distilled from wine or from fermented fruit juice.

Crude fiber—Cellulose, hemicellulose, and similar compounds found in plant tissue.

Distillation—The vaporization of a liquid mixture (water and alcohol; alcohol has a lower boiling point than water) with subsequent collection of the alcohol.

Film yeast—A yeast which grows on the surface of liquids, thereby oxidizing alcohol or lactic acid.

Floaters—Cucumbers and olives filled with gas which cause them to float.

Glucoside—Glucose combined with another compound.

Lactics—The lactic acid bacteria.

Microbial metabolites—Organic compounds produced by metabolism of microorganisms.

Salt stock—Fermented cucumbers preserved by lactic acid and salt.

Sour mash—A mash to which a lactic acid bacterium has been added to lower the pH. A sweet mash does not have these bacteria.

Tannin—A chemical compound found in plants.

REVIEW QUESTIONS

1. What components of cabbage, cucumbers and olives are converted into lactic acid?
2. Which fermentations (kraut, salt stock, cucumbers, or olives) have the most acid?
3. What does salt do in the above fermentations?
4. How many stages are there in the above fermentations?
5. List the dominant bacteria for each stage.
6. List the defects of kraut, salt stock, cucumbers and olives.
7. Outline how beer is produced.
8. What is malt? How is it produced? Why is it needed in fermentations?
9. What are important considerations in mashing?
10. What is the difference in alpha- and beta-amylase?
11. What are defects of beer?
12. How is wine made?
13. What happens during aging of alcoholic beverages?
14. Name two species of yeast that may be used in the production of rum.
15. Define rum, whiskey (rye and bourbon).
16. What is sweet mash? Sour mash?
17. What is a sloe gin? Brandy?
18. What are the two fermentations needed to produce vinegar? Which one is aerobic? Which is anaerobic?
19. What is the Orleans method of producing vinegar? What is the quick process?
20. Why should beer and vinegar be pasteurized?
21. Name three species of yeast used for the production of food yeast.
22. Why are these yeasts used?

REFERENCES

BUCHANAN, R.E. and GIBBONS, N.E. (Editors).   1974.   Bergey's Manual of Determinative Bacteriology, 8th Edition.   Williams & Wilkins Co., Baltimore.

CRUESS, W. V.   1958.   Commercial Fruit and Vegetable Products.   McGraw-Hill Book Co., New York.

DESROSIER, N. W., and DESROSIER, J. N.   1977.   The Technology of Food Preservation, 4th Edition.   AVI Publishing Co., Westport, Conn.

FRAZIER, W. C.   1967.   Food Microbiology.   McGraw-Hill Book Co., New York.

LEUNG, W. W.   1972.   Food Composition Tables for Use in East Asia.   FAO of the UN and USDHEW, Washington, D.C.

PRESCOTT, S. C. and DUNN, C. G.   1949.   Industrial Microbiology.   McGraw-Hill Book Co., New York.

VAUGHN, R. H.   1954A.   Acetic acid-vinegar.   *In* Industrial Fermentations, Vol. 1.   L. A. Underkofler and R. J. Hickey (Editors).   Chemical Publishing Co., New York.

VAUGHN, R. H.   1954B.   Lactic acid fermentation of cucumbers, sauerkraut and olives.   *In* Industrial Fermentations, Vol. 2.   L. A. Underkofler and R. J. Hickey (Editors).   Chemical Publishing Co., New York.

WATT, B. K. and MERRILL, A. L.   1963.   Composition of foods.   USDA Agric. Handbook 8.

15

Microbiology of Milk and Milk Products

MILK AS A SUBSTRATE FOR MICROORGANISMS

Data in Table 15.1 show that milk is an excellent substrate for the growth of bacteria. Milk is a complete medium for growth with its protein (3.5%), carbohydrate (4.7%), minerals and vitamins. The pH range of normal milk is 6.4 to 6.8. The Eh of milk is about 0.30 volts (Sommer and Winder 1970). According to Sommer and Winder, in ordinary mixed fermentation, the lowest Eh value is −0.20 volt. Referring to data in Chap. 5, Eh values of 0.30 volts or 300 mv put milk in an oxidized state in which aerobes can grow easily.

The protein fraction is composed of casein and lactalbumin. Lactalbumin comprises about 0.5% of the total and casein is most of the remaining protein; however, lactoglobulin seldom exceeds 0.1% (Jacobs 1951). Casein is used as a reference protein in determining the quality of other nitrogen sources in feeding rats.

Milk is an emulsion of droplets of fat or globules in water with casein serving as an emulsifying agent. When bacteria grow in milk and produce lactic acid, casein is precipitated out of solution at its isoelectric point (pH 4.6). After this happens, the emulsion is lost.

DOMINANT SPOILAGE FLORA

According to Mossel and Ingram (1955), the dominant spoilage flora under standard conditions of storage are *Streptococcus, Lactobacillus, Microbacterium, Pseudomonas, Flavobacterium* and *Bacillus.*

TABLE 15.1   CHEMICAL COMPOSITION OF COW'S MILK

Component	% or mg/454 g
Protein	3.5%
Fat	3.5%
Carbohydrate	4.9%
Calcium	535 mg
Phosphorus	422 mg
Iron	0.2 mg
Sodium	227 mg
Potassium	654 mg
Thiamin	0.15 mg
Riboflavin	0.78 mg
Niacin	0.3 mg
% acidity of milk[1]	0.1–0.26%
pH of normal cow's milk	6.4–6.8
Eh	0.30 volts

Source: Watt and Merrill (1963).
[1] Acidity consists of casein, albumin, carbon dioxide, citrates and phosphates.

TYPES OF SPOILAGE IN MILK

Data in Table 15.2 show various types of spoilage that may occur in fluid milk. The temperature of storage and pasteurization have an influence upon the types of organisms present. The type of undesirable change may be souring, gas production, proteolysis, ropiness, alkali production, and changes in butterfat and in color.

To prevent deterioration and to extend the shelf-life, milk should be

TABLE 15.2.   TYPES OF SPOILAGE OF MILK

Type of Spoilage	Microflora
Souring	Homofermentative or heterofermentative lactics. At 10°–37°C *Streptococcus lactis* dominant; 37°–50°C *S. thermophilus, S. faecalis* produce about 1% acid followed by lactobacilli to make milk very acid; above 50°C *Lactobacillus thermophilus* produces acid
Gas production	Coliform bacteria, *Clostridium, Bacillus*, yeasts, heterofermentative lactics, and propionics
Proteolysis	*Streptococcus faecalis* var. *liquefaciens, Bacillus cereus, Micrococcus, Pseudomonas, Flavobacterium*
Ropiness	*Alcaligenes, Enterobacter*, lactics
Changes in butterfat	*Pseudomonas, Proteus, Alcaligenes, Bacillus, Micrococcus*
Alkali production	*Pseudomonas fluorescens, Alcaligenes faecalis*
Color changes	*Pseudomonas syncyanea* produces a bluish-gray to brownish color in milk. With *S. lactis* the two produce deep-blue color. *Pseudomonas synxantha* produces yellow color in the cream layer of milk. *Serratia marcescens* produces red milk. *Pseudomonas fluorescens* may cause brown color

Source: Frazier (1967).

stored at 0° to +1.7°C (32° to 35°F). High quality milk which has been pasteurized and stored properly can remain sweet-tasting for several days.

HEAT TREATMENTS

Heat treatments (addition of heat) are given milk to kill pathogens, as in pasteurization or canning. In refrigeration or freezing of dairy products, heat is removed. Table 15.3 summarizes the effect of temperature on the microflora of milk or the milk product. Pasteurization times and temperatures are designed to kill pathogens, not to sterilize. Canning heat treatments are designed to kill spores and to sterilize the product.

Treatments, such as preheating or prewarming are performed prior to another operation such as evaporation or pasteurization. Campbell and Marshall (1975) recommended temperatures of 65°–140°C for a few seconds in the concentration of milk or 10–30 min at 70°–90°C prior to the ultra-high temperature.

These times and temperatures can stimulate germination of bacterial spores, making new vegetative cells very sensitive to heat and easily killed.

PRESERVATION BY FERMENTATION

Several dairy products are preserved by fermentation, resulting in the production of lactic acid. Examples of such foods are cultured buttermilk, Bulgarian milk, yogurt and acidophilus milk (Table 15.4). The acidities produced in these dairy foods are similar to the acidities in fermented cabbage and cucumbers. The lactics are the desirable flora.

Microorganisms are also involved in the manufacture of butter and

TABLE 15.3. HEAT TREATMENTS GIVEN MILK

Treatment	Effect on Microflora
Refrigerator temperature 4°C	Keeps mesophiles in check; on long-term storage, is spoiled by psychrophiles
Pasteurization 62.8°C (145°F) for 30 min or 71.7°C (161°F) for 15 sec	Kills pathogens, yeasts, molds and most vegetative cells
Canning evaporated milk 116°–118°C for 15 min	Kills most spores; depends upon the type and numbers present
Ultra-high temperatures 140°–150°C for 9 sec	Sterilizes product when number of spores is high. Product is canned aseptically

Source: Campbell and Marshall (1975).

TABLE 15.4.   FERMENTATION AS A PRESERVATION METHOD

Type of Product	Flora	Acidity as Lactic (%)
Cultured buttermilk	*Streptococcus cremoris* and *S. lactis*	0.7–0.90
Bulgarian milk	*Lactobacillus bulgaricus*	2.0–4.0[1]
Yogurt	*Lactobacillus bulgaricus* and *Streptococcus thermophilus*	0.85–0.90
Acidophilus milk	*Lactobacillus acidophilus*	1.0–1.5[2]

Source: Campbell and Marshall (1975); Frazier (1967).
[1] Maximum titratable acidity normally produced by this bacterium. Lower acidities may be desirable for taste.
[2] Acidity obtainable in 4 days at 30°C.

cheeses (Table 15.5). Again with respect to the desirable flora, the lactics dominate. *Penicillium roqueforti* and *P. camemberti* are used in the preparation of Roquefort and Camembert cheeses.

If the proper lactics do not dominate or if the proper acidity is not maintained, spoilage of cheeses may occur. This undesirable effect may occur during fermentation, ripening or in the finished cheese (Table 15.6). The result may be gaseous, with unwanted holes, off-flavor, sliminess, bitterness, putrefaction or the growth of mycotoxin-producing fungi on the surface.

Spoilage can be prevented by using "pure" starter cultures and by maintaining strict sanitation in the plant. Sorbic acid can be used to treat the surface of cheeses or the wrappers to prevent the growth of molds.

TABLE 15.5.   MICROORGANISMS USED IN DAIRY PRODUCTS

Product	Microflora
Butter, cultured buttermilk and sour cream	*Streptococcus lactis* and *S. cremoris* for acid production. *Leuconostoc dextranicum, L. citrovorum* and *S. diacetilactis* for flavor and aroma
Cottage cheese, cream cheese, Cheddar, blue, brick cheeses	Mixed lactic culture
Swiss cheese	*Streptococcus thermophilus, Lactobacillus bulgaricus, L. lactis* or *L. helveticus*
Blue cheese, Roquefort, Gorgonzola cheese	*Penicillium roqueforti*
Camembert cheese	*Penicillium camemberti*
Limburger cheese	*Brevibacterium (Arthrobacter) linens*, yeasts and micrococci

Source: Frazier (1967).

TABLE 15.6.   TYPES OF SPOILAGE OF CHEESE

Stage of Manufacture	Flora	Type of Spoilage
During fermentation	Coliforms, lactose-fermenting yeasts, clostridia	Holes in curd and off-flavors
	Bacillus, Leuconostoc, pseudomonads	Holes and bitter flavor, sliminess, off-flavors, gas
During ripening	Lactate-fermenting clostridia, propionic bacteria, heterofermentative lactics, coliforms, micrococci	Gas, bitterness, putrefaction
Finished cheese	Geotrichum lactis, Cladosporium, Penicillium, film yeast	Surface growth of molds and yeasts

Source: Frazier (1967).

GLOSSARY

Evaporated milk—Evaporated milk is condensed by the evaporation under vacuum of a considerable portion of the water. The milk is canned.

Starter cultures—Cultures of lactic acid bacteria (or any other microorganism) that are used to ensure the success of the fermentation.

REVIEW QUESTIONS

1. List the components of milk and state why milk is a good medium for bacteria.
2. List various types of spoilage of milk and give the type of microflora using it.
3. What type of heat treatment is given milk to extend its shelf-life?
4. What is the dominant spoilage flora of milk?
5. What are some foods in which lactic acid aids in its preservation?
6. What are desirable microorganisms used in the manufacture of butter, cottage cheese, Swiss cheese, Roquefort cheese, Limburger cheese, cultured buttermilk, Bulgarian milk, acidophilus milk and yogurt?

REFERENCES

CAMPBELL, J. and MARSHALL, R. T. 1975. The Science of Providing Milk for Man. McGraw-Hill Book Co., New York.

FRAZIER, W. C. 1967. Food Microbiology. McGraw-Hill Book Co., New York.

JACOBS, M. G. 1951. Milk, cream and dairy products. In The Chemistry and Technology of Food and Food Products. Interscience Publishers, New York.

MOSSEL, D. A. A. and INGRAM, M.   1955.   The physiology of the microbial spoilage of foods.   J. Appl. Bacteriol. *18,* 232–268.

SOMMER, H. H. and WINDER, W. C.   1970.   The physical chemistry of dairy and food products.   Dep. Food Sci., Univ. Wisconsin.

WATT, B. K. and MERRILL, A. L.   1963.   Composition of foods.   USDA Agric. Handbook *8.*

16

Microbiology of Refrigerated and Frozen Foods

REFRIGERATED FOODS

Control of Microflora and Enzymes of Foods

Temperature. Refrigerated foods are those foods which are kept above their freezing points to extend their shelf-life. Desrosier and Desrosier (1977) listed perishable foods as foods being stored as low as −1.9°C (30°F) for cream cheese to 19.5°C (60°F) for mature green pineapple. Most fruits are stored at −0.6 to 0°C (31° to 32°F) except for lemons, cranberries, limes, mangoes, papayas and ripe pineapples which are stored at higher temperatures. Most vegetables are also stored at 0°C (32°F). Cucumbers, eggplants, muskmelons, honeydew, okra, olives, potatoes, pumpkins, squash and tomatoes are the exceptions which are held at higher storage temperatures (Wright *et al.* 1954).

Fruits and vegetables are unique foods since they are still living and respiring. Respiration involves the giving off of heat so that more refrigeration is needed to store these foods than the same mass of meat, milk, eggs or fish. The lower the temperature at which these foods can be stored without injury, the more the rate of enzyme action can be slowed to prevent their catabolic reactions and to preserve the freshness of the food.

As was discussed in Chap. 5, microorganisms can be classified according to their growth temperature. Psychrophiles can grow at temperatures from −7° to 10°C and mesophiles from 10° to 45°C. Salle (1973) used the term psychrophile for organisms which grow at 5°C or below. In this chapter, psychrophilic microflora will be considered to be those microorgansims which grow from −7°C to 10°C.

By definition of psychrophiles and mesophiles, multiplication of mesophilic microflora is restricted at cold storage or refrigerated temperatures, and psychrophiles will thus dominate in the spoilage of foods stored for a considerable time.

Food poisoning bacteria can be controlled by the storage temperature with the exception of *Clostridium botulinum* type E. (See Table 16.1.) Schmidt *et al.* (1961) observed that type E can grow at 3.3°C (37°F). This food poisoning bacterium may be a problem if foods are stored at a temperature above 3.3°C for extended periods of time.

The lowest temperatures supporting growth of other food poisoning and infection bacteria are given in Table 16.1. Obviously, the growth of these organisms can be controlled if the temperature is lowered to 2°C (36°F). The question is how efficient the refrigeration system is in preventing the temperature from rising into the growth range and/or from dropping to a temperature where freezing may occur and cause freeze damage to the food.

Humidity and Air Circulation. To maintain temperature and humidity within a refrigerated room or a refrigerator, the air must be circulated. Also, to prevent the drying of foods like fruits and vegetables a humidity of about 90–95% (varies with foodstuff) is needed. As we discussed in Chap. 5, microorganisms need water to reproduce. If too high a humidity is maintained, more microorganisms will grow even at the reduced temperature. The adjustment of the relative humidity is a trade-off between microbial growth and moisture loss from the product.

Ultraviolet Lights. Ultraviolet lights may be used to prevent the growth of microorganisms on the product during refrigerator storage. An example is to keep the microflora at low levels during aging of beef to make it more tender. Because UV has low penetrating power, its use is limited to surfaces.

Gaseous Atmosphere. Certain foods may be stored in gaseous atmospheres to extend their shelf-life while being held at low temperatures. Apples may be stored under certain levels of carbon dioxide. This would tend to prevent the growth of aerobic microorganisms such as filamentous fungi. Carbon dioxide also has been used in the shipment of beef to prevent the growth of aerobes.

TABLE 16.1.   LOWEST TEMPERATURES FOR GROWTH OF FOOD POISONING AND FOOD INFECTION BACTERIA

Bacterium	Lowest Temp of Growth (°C)	Reference
Staphyloccus aureus	6.7	Angelotti *et al.* (1961)
Clostridium botulinum		
Type A and B	12.5	Ohye and Scott (1953)
Type E	3.3	Schmidt *et al.* (1961)
Salmonella heidelberg	5.3	Matches and Liston (1968)
S. typhimurium	6.2	Matches and Liston (1968)
Bacillus cereus	10	Gordon *et al.* (1973)
Clostridium perfringens	20	Smith and Hobbs (1974)

Type of Spoilage

Nature of Psychrophiles. The unique properties of psychrophilic bacteria allow them to grow at lower temperature than do the mesophiles. These are outlined in Table 16.2. Because of these unique properties and possibly others, these microflora spoil foods stored above their freezing points but at refrigerator temperatures.

Psychrophilic fungi also exist. Gunderson (1961) isolated and identified 113 molds from frozen convenience food products. Fifty-two molds, or 46%, grew at 5°C or below. According to Gunderson, fluctuating temperatures allowed moisture to migrate within the product, thus providing the necessary water for mold growth.

Some of the fungi with which we are familiar are shown in Table 16.3. The fungus which Gunderson found as the single most-encountered psychrophilic fungus was *Pullularia pullulans* or *Aureobasidium pullulans*. This fungus is usually not listed in food microbiology texts.

Because the moisture level may be low, molds are more apt to dominate during fluctuating temperatures of the foods than bacteria. The fungi that Gunderson (1961) found associated with certain types of foods are listed in Table 16.4. Here, again, all the microorganisms are fungi.

EFFECT OF PRETREATMENTS PRIOR TO FREEZING

Blanching Vegetables. The primary reason for blanching prior to freezing is to inactivate the enzymes. This process would also reduce the microflora.

TABLE 16.2.   UNIQUE PROPERTIES OF PSYCHROPHILES AS COMPARED TO MESOPHILES

Property	Psychrophile	Mesophile
Unsaturated fatty acids	Greater proportion in cells	Less unsaturated in cells
Polysaccharides	Greater synthesis	Less synthesis
Pigments (carotenoid)	Greater synthesis under psychrophilic conditions	Less synthesis under mesophilic conditions
Products of fermentation	Acid plus gas under psychrophilic conditions for some strains	Acid only
Metabolic rate	Slower	Faster
Transport of solutes across membranes	More transparent	Less transparent
Cell size	Some are larger	Some are smaller
Producers of flagella	More efficient producers	Less efficient producers
Respond to aeration	Respond more favorably	Respond less favorably

Source: Jay (1970).

TABLE 16.3.   PSYCHROPHILIC STRAINS[1] OF COMMON MESOPHILIC FUNGI ASSOCIATED WITH FOOD SPOILAGE

Genus	No. of Species	Frequency of Encounter
Alternaria	2	
Botrytis cinerea	1	2
Cladosporium	3	
Geotrichum candidum	1	4
Mucor	3	
Rhizopus nigricans	1	
Penicillium	22	
Pullularia pullulans	1	1[2]
Phoma sp.	1	3

Source: Gunderson (1961).
[1] Grew at 5°C or lower.
[2] "1" means it was the most frequently encountered mold in a frozen product.

TABLE 16.4.   TYPES OF FUNGI ASSOCIATED WITH CERTAIN TYPES OF FOODS

Type of Food	Microorganism
Chicken pot pies	*Aleurisma carnis*
Cherry fruit fillings	*Phoma* sp.
Blueberry fruit fillings	*Pullularia pullulans*
	Penicillium lapidosum
	Penicillium thomii

Source: Gunderson (1961).

Peeling of Fruits or Vegetables. Vegetables may be lye-peeled. Such a treatment would lower the surface contaminants.

Chlorination. In-plant chlorination of fruit, vegetable, fish and meat products would lower the microbial population.

Rapid Handling. Rapid handling prior to freezing would keep the growth curve in the lag phase, thereby keeping the numbers of microorganisms low.

EFFECT OF FREEZING, STORAGE AND THAWING ON SURVIVAL OF MICROORGANISMS

Cooling Rates

The more rapid the cooling rate, the greater the numbers of microorganisms surviving. Borgstrum (1961) suggested several times and temperatures for slow, rapid and ultra-rapid cooling rates. He considered, for example, 1°C in 50 min to be slow, 1°C in 1 min to be rapid, and 5°C in 1 sec to be ultra-rapid. Ultra-rapid cooling rates are used to preserve microorganisms.

Most foods would be frozen slowly at home or rapidly by commercial methods. With the cost of energy and the availability of energy for freezing foods, these rates will probably continue.

If a food is frozen slowly, the pure water freezes first, leaving a more concentrated substance. (Salts and sugars would depress the freezing points.) This would continue until all the material is frozen.

Exposure to Electrolytes and Osmotic Pressure

As a food (orange juice, for example) is frozen the microflora are exposed to a more concentrated solution. The resulting high osmotic pressure and high level of electrolytes may be lethal or damaging to the microbial cell, thus causing death of the cell. Rapid freezing would minimize these harmful effects.

During the freezing process, some of the bacteria die. The exposure time to adverse conditions will determine the number of survivors. Freezing, however, does not sterilize the food.

Storage Time

Depending upon the storage time and temperature, the number of bacteria decline in numbers. According to Borgstrom (1961), more microorganisms are destroyed at −4° than at −15° or −24°C. The rate of reduction during storage is much less than during the freezing phase. Sporeformers survive better than nonsporeformers. Enterococci survive longer than coliforms.

Thawing Period

During thawing, the surviving cells are subjected to the same kind of stresses as they were during freezing (electrolytes and osmotic pressure effects).

Holding Before Use

Frozen vegetables should be cooked while still frozen. If allowed to thaw and then held, growth of the survivors may be rapid, resulting in spoilage.

GLOSSARY

Electrolyte—A substance which conducts electricity. Examples are NaCl, MgSO$_4$, etc. Salts are electrolytes.

Frozen foods—Foods that are stored at temperatures below their freezing points.

Refrigerated foods—Foods stored at temperatures above their freezing point.

REVIEW QUESTIONS

1. Why do we refrigerate foods?
2. Why do we not refrigerate all foods at the same temperature?
3. What makes fruits and vegetables unique among refrigerated foods?
4. Based upon their temperature of growth, how may we classify bacteria?
5. List the lowest temperatures for growth of food poisoning and food infection bacteria.
6. List ways to reduce the microflora of foods held at refrigeration temperatures.
7. How are psychrophiles different from mesophiles?
8. List some psychrophilic fungi which spoil foods.
9. What pretreatments given vegetables prior to freezing influence the microflora?
10. Give the effect of freezing, holding at subzero temperatures, and thawing on microorganisms.

REFERENCES

ANGELOTTI, R., FOSTER, M. J. and LEWIS, K. H. 1961. Time-temperature effects on salmonellae and staphylococci in foods. I. Behavior in refrigerated foods. II. Behavior at warm holding temperatures. Am. J. Public Health *51,* 76–88.

BORGSTROM, G. 1961. Unsolved problems in frozen food microbiology. *In* Proc. Low Temperature Microbiology Symp. Campbell Soup Co., Camden, N.J.

CRISLEY, F. D. and HELZ, G.E. 1961. Some observations of the effect of filtrates of several representative concomitant bacteria on *Clostridium botulinum* type A. Can. J. Microbiol. *7,* 633–639.

DESROSIER, N. W. and DESROSIER, J. N.   1977.   The Technology of Food Preservation, 4th Edition.   AVI Publishing Co., Westport, Conn.

GORDON, R. E., HAYNES, W. C. and PANG, C.   1973.   The genus *Bacillus.* USDA Agric. Handbook *427.*

GUNDERSON, M. F.   1961.   Mold problem in frozen foods.   *In* Proc. Low Temperature Microbiology Symp.   Campbell Soup Co., Camden, N.J.

JAY, J. M.   1970.   Modern Food Microbiology.   Van Nostrand Reinhold Co., New York.

MATCHES, J. R. and LISTON, J.   1968.   Low temperature growth of *Salmonella.* J. Food Sci. *33,* 641–645.

OHYE, D. F. and SCOTT, W. J.   1953.   The temperature relations of *Clostridium botulinum* types A and B.   Aust. J. Biol. Sci. *6,* 178–189.

SALLE, A. J.   1973.   Fundamental Principles of Bacteriology.   McGraw-Hill Book Co., New York.

SCHMIDT, C. F., LECHOWICH, R. V. and FOLINAZZO, J. F.   1961.   Growth and toxin production of type E *Clostridium botulinum* below 40°F.   J. Food Sci. *26,* 626–630.

SMITH, L. D. S. and HOBBS, G.   1974.   *Clostridium. In* Bergey's Determinative Bacteriology.   R. E. Buchanan and N. E. Gibbons (Editors).   Williams & Wilkins Co., Baltimore.

WRIGHT, R. C., ROSE, H. and WHITEMAN, T. M.   1954.   The commercial storage of fruits, vegetables, and florist and nursery stocks.   USDA Agric. Handbook *66.*

17

Food Poisoning and Contamination of Foods with Mycotoxins

FOOD POISONING

Clostridium perfringens

Symptoms. *C. perfringens* causes a rather mild food poisoning. Symptoms are abdominal pain, diarrhea, nausea and vomiting. Fever, shivering and headaches are rarer. Usually, the symptoms which last only a day begin about 12 hr (8–22 hr may be the range) after ingestion.

Foods Involved. The Center for Disease Control in Atlanta, Georgia issues annual summaries of *Foodborne Outbreaks*. The data in Table 17.1 were constructed from these reports. In 29 of the confirmed cases reported from 1969–1972 (Table 17.1), only three outbreaks were associated with nonmeat products (salad and/or potato, rice, and macaroni and cheese). Eleven outbreaks were associated with poultry products.

Prevention. *C. perfringens* is a sporeforming anaerobic bacterium so it is not realistic to consider foods free of spores. The Food and Drug Administration considers the following as means of controlling *C. perfringens* food poisonings:

(a) Strict adherence to good sanitary practices in the handling, preparation and serving of foods, especially meats, gravies and meat dishes.

(b) Meats should be properly cooked, held hot above 52°C (125°F) and served hot.

(c) If cooked for later use, meats should be cooled rapidly in small lots in a refrigerator to 7°C (45°F) or below.

(d) Leftover meat or meat dishes should receive an adequate heat treatment before being used, and leftover gravy should be

TABLE 17.1.   FOOD ASSOCIATED WITH CONFIRMED CASES OF FOOD POISONING
CAUSED BY *C. PERFRINGENS*

Year	Food	Location
1969	Pork	School
	Chicken gravy	Restaurant
	Chicken	Home
	Beef stew	Hospital
	Turkey	Cafeteria-catered
	Beef	Home
	Turkey	Home-catered
	Salad and/or potato	Restaurant
	Burritos	School
	Braised beef or rice	School
	Roast beef	Restaurant
	Turkey and/or gravy	Hospital
	Turkey	School
	Macaroni and cheese	Cafeteria
	Rice	Restaurant
1970	Beef sandwich	Restaurant
	Barbecue	Picnic
	Beef roast	Restaurant
	Hamburger	School
	Turkey	Cafeteria
1971	All listed were probable *C. perfringens*	
1972	Turkey	1
	Meat sauce	1
	Chicken	1
	Gravy	1
	Beef	1
	Roast beef	1
	Chicken casserole	1
	Turkey	1
	Meat sauce	1

Source: CDC (1969–1972).
[1] Not listed in this issue.

brought to a rolling boil before being placed back in the steam
table.

(e)   Cold cuts and sliced meats should be maintained cold, below 7°C
(45°F), and served cold, not at room temperature.

Media and Methods of Detection. Media and methods used in detecting *C. perfringens* are listed in Table 17.2. These characteristics of the
species are described in *Bergey's Manual of Determinative Bacteriology*
(Breed and Murray 1957; Buchanan and Gibbons 1974).

One medium for the detection or enumeration of *C. perfringens* is the
sulfite-polymyxin-sulfadiazine medium which consists of ingredients listed
in Table 17.3. SFP medium, given in Table 17.4, may also be used. Both
substrates are similar in that they contain a source of iron and sulfite. It
is characteristic of *C. perfringens* to reduce sulfite to sulfide with the
consequent blackening of media containing iron to form iron sulfide (Smith
1955).

TABLE 17.2.   GROWTH CHARACTERISTICS OF *C. PERFRINGENS*

Characteristic	Reference
Anaerobic, non-motile rods; Gram-positive	Breed *et al.* (1957); Buchanan and Gibbons (1974)
Liquefaction and blackening of gelatin	Breed *et al.* (1957)
Acid coagulation of litmus milk. Clot torn with profuse gas production but no digestion	Breed *et al.* (1957)
Nitrites produced from nitrates	Breed *et al.* (1957)
Nitrate reduction is variable depending upon the basal medium	Buchanan and Gibbons (1974)
Optimum growth temperature 45°C; range 20°–50°C	Buchanan and Gibbons (1974)
Lecithinase production by growing on egg yolk agar, producing an opaque precipitate	Breed *et al.* (1957)

TABLE 17.3.   COMPONENTS OF SULFITE-POLYMYXIN-SULFADIAZINE MEDIUM (SPS)[1]

Ingredient	g/liter	Function of Ingredient
Tryptone	15.0	Provide C and N
Yeast extract	10.0	Provide B vitamins
Ferric citrate	0.5	Provide a source of Fe
Sodium sulfite	0.5	Provide a source of sulfite to be reduced to sulfide
Sodium thioglycollate	0.1	Lower the Eh of the medium
Tween 80	0.05	Function as a surface active agent
Sodium sulfadiazine	0.12 ⎫	To inhibit Gram-negative
Polymyxin B sulfate	0.01 ⎭	bacteria and Gram-positive cocci
Agar	15.0	To make medium solidify

[1] Product of Difco Laboratories, Detroit, Mich. (Code 0845).

TABLE 17.4.   COMPONENTS OF SFP[1] MEDIUM FOR THE DETECTION AND ENUMERATION OF *C. PERFRINGENS* IN FOOD

Ingredient	g/liter	Function of Ingredient
Tryptose	15	Provide C and N
Yeast extract	5	Provide B vitamins
Soytone	5	Provide C and N
Ferric ammonium citrate	1	Provide source of Fe
Sodium bisulfite	1	Provide source of sulfite
Agar	20	To make medium solidify
Polymyxin B kanamycin (1 vial)	0.012	To inhibit Gram-negative bacteria
Egg yolk 50% in saline, 100 ml		To demonstrate lecithinase production

[1] Product of Difco Laboratories, Detroit, Mich. (Code 0811).

According to Smith (1955), morphology, colony shape, and reaction on egg-yolk agar should be used for tentative identification of *C. perfringens*. Lewis and Angelotti (1964) used motility-nitrate medium and sporulation broth to confirm the black colonies from the SPS agar.

In the analysis of foods by the Division of Microbiology (FDA 1976), Harmon reported that *C. perfringens* cells (Gram-positive, nonmotile, sporeforming, obligate anaerobes) reduce nitrate to nitrite and produce stormy fermentation in iron milk. The media used in studying this anaerobe are listed in Table 17.5. See Speck (1976) for media and additional methods for studying *C. perfringens*.

Investigation of a Foodborne Outbreak. Both epidemiological and laboratory data should be collected regardless of the type of microorganism. For example, the same form would be used whether the poisoning is caused by staphylococci or by salmonellae. An investigation of a foodborne outbreak is recorded on the form shown in Fig. 17.1. The correlation of clinic and laboratory data necessary to show cause of an outbreak is as outlined in Table 17.6.

TABLE 17.5. MEDIA USED IN STUDYING *C. PERFRINGENS*

Medium	Use
Chopped liver broth or cooked meat medium	General cultivation
Indole-nitrate medium	To test for production of indole and reduction of nitrate to nitrite
Fluid thioglycollate medium with Eh indicator and glucose	Cultivate the bacterium, enrichment
Iron milk medium	To determine stormy fermentation, coagulation, digestion and blackening of milk
Liver-veal-egg yolk agar	To determine the production of lecithinase zones (opalescent zone 2–5 mm in diameter around each colony)

Clostridium botulinum[1]

Symptoms. The symptoms of botulism are fatigue, dizziness, headache, nausea, vomiting, diarrhea, double vision, dryness of mouth, constriction of the throat, a swollen and coated tongue, and normal to subnormal temperature. Paralysis spreads to the respiratory system and heart. Death usually comes at 3 to 6 days. The onset of symptoms varies from 8 to 28 hr.

Foods Involved. During the period of 1899–1973, the Center for Disease Control (CDC) has recorded the data listed in Table 17.7. The data show that the leading cause of botulism in the United States was toxin type A. The principal kind of food involved was vegetables.

The data from CDC with the specific food serving as the vector in cases of botulism from 1950–1973 are found in Table 17.8. During this period, there was a total of 211 outbreaks; but the data in Table 17.8 cover only the 121 outbreaks in which the specific food was mentioned. The data in Table 17.8 for the period 1950–1973 show the same trend for types A and B as

---

[1] See Smith (1977) for a description of the organism, its toxins, and the disease.

TABLE 17.6.   CENTER FOR DISEASE CONTROL GUIDELINES FOR CONFIRMATION ON FOODBORNE OUTBREAKS

Bacterium	Clinical Syndrome	Laboratory Criteria
C. perfringens	Incubation period 8–22 hr Lower intestinal syndrome (majority of cases with diarrhea with little vomiting or fever	Organisms of same serotype in epidemiologically incriminated food and stool of ill individuals Isolation of organisms with same serotype in stool of most ill individuals $>10^5$ bacteria in epidemiologically incriminated food provided specimen properly handled
C. botulinum	Clinical syndrome compatible with botulism (See CDC Botulism Manual)	Food epidemiologically incriminated or Detection of botulinal toxin in human sera, feces, or food or Isolation of C. botulinum bacteria from food
B. cereus	Incubation period 1–16 hr Gastrointestinal syndrome	Isolation of $>10^5$ bacteria in epidemiologically incriminated food or Isolation of bacteria in stools of ill person

Source: CDC (1972).

found in the period 1899–1973. Type A caused the most cases, followed by B. During the period 1950–1973, however, there were many outbreaks due to E. Only one outbreak of type F occurred during this time. This was in homemade venison jerky.

If one examines the table (17.8) as to the source of botulism, we find that only 13 outbreaks were attributed to commercially prepared food whereas 108 (89.2%) were caused by either home canned foods or home preserved foods. Obviously, the homemaker is not doing as good a job in preserving foods.

The CDC handbook covering the period of 1950–1973 does not tell all the specifics of the cases of botulism. We can learn something, however, by grouping the foods as in Table 17.9 which is a summary of Table 17.8.

Let us first consider the acid foods in which botulism occurred. These outbreaks are less common than for low acid foods. The minimum pH permitting the outgrowth of type A and B spores is 4.8. Furthermore, the heat resistance of A and B spores is lower in acid foods. The pH for tomatoes is usually 4.1–4.4; for huckleberries, 2.8–2.9; for peaches, 3.4–4.2; for tomato juice 3.9–4.4; for blackberries, 3.0–4.2; for pickles, 2.6–3.8; for apple butter,

FORM APPROVED
OMB NO. 68-R557

E.   INVESTIGATION OF A FOODBORNE OUTBREAK

1. Where did the outbreak occur?

   State _____ (1,2) City or Town _____ County _____

2. Date of outbreak: (Date of onset 1st case)

   _____ (3-8)

3. Indicate actual (a) or estimated (e) numbers:

   Persons exposed _____ (9-11)

   Persons ill _____ (12-14)

   Hospitalized _____ (15-16)

   Fatal cases _____ (17)

4. History of Exposed Persons:

   No. histories obtained _____ (18-20)

   No. persons with symptoms _____ (21-23)

   Nausea _____ (24-26) Diarrhea _____ (33-35)

   Vomiting _____ (27-29) Fever _____ (36-38)

   Cramps _____ (30-32) Other, specify _____ (39)

5. Incubation period (hours):

   Shortest _____ (40-42) Longest _____ (43-45)

   Approx. for majority _____ (46-48)

6. Duration of Illness (hours):

   Shortest _____ (49-51) Longest _____ (52-54)

   Approx. for majority _____ (55-57)

7. Food-specific attack rates: (58)

Food Items Served	Number of persons who ATE specified food				Number who did NOT eat specified food			
	Ill	Not Ill	Total	Percent Ill	Ill	Not Ill	Total	Percent Ill

FIG. 17.1.   REPORT FORM FOR FOODBORNE OUTBREAK

Reprinted with permission of U.S. Department of Health, Education and Welfare, Public Health Service, Center for Disease Control, Bureau of Epidemiology.

8. Vehicle responsible (food item incriminated by epidemiological evidence): (59,60)

9. Manner in which incriminated food was marketed: (Check all applicable)

(a) Food Industry   (61)   (c) Not wrapped 1 (63)

Raw 1   Ordinary Wrapping ... 2

Processed 2   Canned........... 3

Home Produced   Canned–Vacuum Sealed.. 4

Raw 3   Other (specify) 5

Processed 4

(b) Vending Machine.... 1 (62)   (d) Room Temperature 1 (64)

                      Refrigerated.......... 2

                      Frozen............. 3

                      Heated............. 4

If a commercial product, indicate brand name and lot number

10. Place of Preparation of Contaminated Item: (65)

Restaurant 1

Delicatessen 2

Cafeteria 3

Private Home 4

Caterer 5

Institution:

School 6

Church 7

Camp 8

Other, specify ... 9

11. Place where eaten: (66)

Restaurant......... 1

Delicatessen 2

Cafeteria.......... 3

Private Home 4

Picnic 5

Institution:

School......... 6

Church......... 7

Camp 8

Other, specify ... 9

DEPARTMENT OF HEALTH, EDUCATION, AND WELFARE
PUBLIC HEALTH SERVICE
CENTER FOR DISEASE CONTROL
BUREAU OF EPIDEMIOLOGY
ATLANTA, GEORGIA 30333

LABORATORY FINDINGS (Include Negative Results)

12. Food specimens examined: (67)

Specify by "X" whether food examined was original (eaten at time of outbreak) or check-up (prepared in similar manner but not involved in outbreak)

Item	Orig.	Check up	Findings Qualitative	Findings Quantitative
Example: beef	X		C. perfringens, Hobbs type 10	2×10^6/gm

13. Environmental specimens examined: (68)

Item	Findings
Example: meat grinder	C. perfringens, Hobbs Type 10

14. Specimens from patients examined (stool, vomitus, etc.): (69)

Item	No. Persons	Findings
Example: stool	11	C. perfringens, Hobbs Type 10

15. Specimens from food handlers (stool, lesions, etc.): (70)

Item	Findings
Example: lesion	C. perfringens, Hobbs type 10

16. Factors contributing to outbreak (check all applicable):

	Yes	No
1. Improper storage or holding temperature	1 ☐	2 ☐ (71)
2. Inadequate cooking	1 ☐	2 ☐ (72)
3. Contaminated equipment or working surfaces	1 ☐	2 ☐ (73)
4. Food obtained from unsafe source	1 ☐	2 ☐ (74)
5. Poor personal hygiene of food handler	1 ☐	2 ☐ (75)
6. Other, specify	1 ☐	2 ☐ (76)

FIG. 17.1. (Continued)

17. Etiology: (77, 78)

Pathogen _____

Chemical _____

Other _____

Suspected . . . . ☐ 1 (79)

Confirmed . . . . ☐ 2

Unknown . . . . ☐ 3

18. Remarks: Briefly describe aspects of the investigation not covered above, such as unusual age or sex distribution; unusual circumstances leading to contamination of food, water; epidemic curve; etc. (Attach additional page if necessary)

Name of reporting agency: (80)

Investigating official:	Date of investigation:

NOTE: Epidemic and Laboratory Assistance for the investigation of a foodborne outbreak is available upon request by the State Health Department to the Center for Disease Control, Atlanta, Georgia 30333.

To improve national surveillance, please send a copy of this report to:

Center for Disease Control

Attn: Enteric Diseases Section, Bacterial Diseases Branch

Bureau of Epidemiology

Atlanta, Georgia 30333

Submitted copies should include as much information as possible, but the completion of every item is not required.

CDC 4.245 (BACK)

1-74

TABLE 17.7. FOOD PRODUCTS CAUSING BOTULISM OUTBREAKS 1899–1973

Botulinum Toxin Type	Vege- tables	Fish & Fish Products	Fruits	Condi- ments	Beef	Milk & Milk Products	Pork	Poultry	Other	Total
A	96	7	22	16	3	2	2	1	5	154
B	24	3	5	4	1	2	1	2	0	42
E	1	19								20
F					1					1
A & B	2									2
Total	123	29	27	20	5	4	3	3	5	219

Source: CDC (1974).

TABLE 17.8. FOODBORNE BOTULISM OUTBREAKS 1950–1973
(Only outbreaks where food was listed by CDC were used in this table)

Year	Toxin Type	Food	Source
1950	E	Beluga	Home preserved
1951	NA	Tomatoes	Home canned
	NA	String beans	Home canned
	A	Greens	Home canned
	A	Asparagus	Home canned
1952	E	Beluga flippers	Home preserved
	NA	Olives	Home canned
	A	Mushrooms	Home canned
	NA	Mushrooms	Home canned
1953	A	Huckleberry juice	Home canned
	B	String beans	Home canned
	NA	Beets	Home canned
	NA	Cheese	Homemade
	NA	Frozen lobster tail	Commercially processed
1954	A	Peaches	Home canned
	A	Okra	Home canned
	NA	Asparagus	Home canned
	NA	Beets	Home canned
	A	Beets	Home canned
1955	B	Green olives	Home canned
	A	Chili peppers	Home canned
	A	Spinach	Home canned
	NA	Mushrooms	Home canned
1956	E	Beluga	Home preserved
	E	Ougruk	Home preserved
	A	Olives	Home canned
	NA	Potatoes	Home canned
	NA	Pickled pigs' feet	Home canned
	NA	Beet greens	Home canned
	B	Swiss chard	Home preserved
1957	A	Tuna fish	Home canned
	NA	Green beans	Home canned
	NA	String beans	Home canned
	NA	Mushrooms	Home canned
	NA	Sausage	Homemade
	NA	Gluten	Home canned
1958	NA	Mushrooms	Home canned
	NA	Beans	Home canned
1959	E	Fish eggs	Home preserved
	E	Seal or whale flippers	Home preserved
	E	Fish eggs	Home preserved
	NA	String beans	Home canned
	NA	Mushrooms	Home canned
	A	Corn and chicken mash	Home canned
	NA	Beans	Home canned
	A	Green beans	Home canned
	NA	Beets	Home canned
	NA	Beets	Home canned
1960	E	Salmon eggs	Home preserved
	NA	Beets	Home canned
	NA	Green beans	Home canned

TABLE 17.8. (*Continued*)

Year	Toxin Type	Food	Source
	NA	Beets	Home canned
	E	Smoked fish	Commercially processed
	NA	Frozen chicken pie	Commercially processed
1961	E	Salmon eggs	Home processed
	NA	Chili	Home processed
1962	NA	Green beans	Home canned
	NA	Mushrooms	Home canned
	NA	Red peppers	Home canned
	A	Chili	Home canned
	NA	Corn	Home canned
1963	E	Smoked whitefish	Commercially packaged
	NA	Mushrooms	Home canned
	NA	Figs	Home canned
	A	Chili peppers	Home canned
	A	Green beans	Home canned
	B	Corn	Home canned
	E	Whitefish	Commercially processed
	E	Tuna fish	Commercially canned
	E	Smoked whitefish	Commercially processed
	A	Liver paste	Commercially canned
	B	String beans	Home canned
	B	Green beans	Home canned
1964	NA	Chili beans or green peppers	Home canned
	NA	Peppers	Home canned
	A	Pickles	Home canned
	B	Green beans	Home canned
1965	NA	Tomato juice	Home canned
	A	Tuna	Home processed
1966	F	Venison jerky	Homemade
	A	Beets	Home canned
	B	Ham from Germany	Commercially packed
	B	Mushrooms	Home canned
1967	E	Seal flippers	Home prepared
	NA	Green beans	Home canned
	E	Whitefish	Commercial source but home canned
	B	Peppers	Home canned
1968	NA	Fish	Home processed
	NA	Vegetables	Home canned
	NA	Fruit preserves	Home prepared
	A	Chicken livers	Commercially processed
	NA	Hamburger	Home prepared
	B	Chicken soup	Home processed
1969	A	Potato salad	Home prepared
	NA	Apple butter	Home canned
	A	Mushrooms	Home canned
	B	Tomato juice	Home canned
	A	Pumpkin pie	Home canned
	NA	Beets and/or carrots, onions and cauliflower	Home canned

TABLE 17.8. (*Continued*)

Year	Toxin Type	Food	Sources
1970	E	Fermented whale meat, blubber, skin	Home processed
	A	Olives	Home canned
	A	Chili peppers	Home canned
	A	Spaghetti sauce	Home canned
	A	Bean juice	Home canned
1971	E	Frozen whitefish	Home processed
	A	Chili peppers	Home canned
	B	Fermented soybeans	Home processed
	NA	Celery	Home processed
	A	Antipasto	Home processed
	A	Vichyssoise soup	Commercially canned
	B	Peppers	Home canned
1972	A	Peppers	Home canned
	NA	Vegetables	Home canned
1973	B	Smoked whitefish	Home preserved
	A	Chili sauce	Home canned
	A	Smoked salmon	Home canned
	B	Green beans	Home canned
	B	Blackberries	Home canned
	B	Peppers	Commercially canned
	E	Salmon eggs	Home processed
	A	Smoked salmon	Home processed

Total 121	Type E 19
Type A 35	Type F 1
Type B 18	NA (Information not available) 48

Source: CDC (1974).

probably near that of apple or 3.4–3.5; the pH of fruit preserves was probably acid, too.

All of these foods were home-canned. Of course, there are all kinds of possibilities if wrong techniques were used in canning. This is usually the case. For example, if mold grew in any of the jars and used the acid of the food product, the pH could rise higher than 4.8, permitting the spores to germinate and to produce the toxin. There should be no growth of *C. botulinum* in canned food if the food has been heated correctly and if adequate containers have been used.

The low-acid foods have been a problem over the years. This is why the National Canners Association has spent so much time in developing adequate processes for these foods. If one considers the low incidence of botulism in the billions and billions of cans commercially processed, we can see what an excellent job industry has done. Ideally, we want no cases of botulism either in home- or commercially-processed foods. Home canners have done the poorest job because they have not followed the basic principles of canning. The foods listed in the low-acid list in Table 17.9 have

TABLE 17.9.   OUTBREAKS OF BOTULISM CAUSED BY CANNED FOODS, MEAT AND MEAT PRODUCTS, FISH AND FISH PRODUCTS, AND MISCELLANEOUS FOODS

Canned Foods	Meat and Meat Products	Fish and Fish Products	Miscellaneous Foods
Acid Foods			
Tomatoes	Sausage	Fish eggs	Seal or
Huckleberry juice	Ham	Salmon eggs	whale flippers
Peaches	Hamburger	Smoked fish	Fermented
Tomato juice	Pickled pigs' feet	Smoked whitefish	whale meat
Blackberries	Venison jerky	Tuna	Beluga
Pickles	Liver paste	Smoked salmon	Ougruk
Fruit preserves	Chicken livers	Frozen whitefish	Gluten
Apple butter		Frozen lobster	Fermented
		tails	soybeans
Low-Acid Foods			Celery
String beans			Antipasto
Greens			Swiss chard
Asparagus			Potato salad
Olives			Cheese
Mushrooms			
Beets			
Okra			
Chili peppers			
Spinach			
Green olives			
Potatoes			
Beet greens			
Tuna fish			
Beans			
Corn and chicken mash			
Red peppers			
Pumpkin			
Onions and cauliflower			
Potato soup			
Vegetables			
Figs			

Source: CDC (1974).

pH values greater than 4.8. All these foods have to be processed at 115.6°C (240°F) to kill the spores of *C. botulinum*. If enough heat is not applied or if wrong types of containers which allow leakage are used, botulism may occur when the food is eaten after toxin has been produced.

Sausage, ham, hamburger, pickled pigs' feet and venison jerky were the meat products involved in botulism.

Type E toxin was found in fish eggs, salmon eggs, smoked fish, smoked whitefish, smoked salmon and tuna. Type E is heat labile as compared to A and B.

Seal or whale flippers, fermented whale meat, beluga, ougruk, gluten, fermented soybeans, celery, antipasto, Swiss chard, potato salad and cheese were among the miscellaneous foods involved in botulism during the 1950–1973 period.

Prevention. The obvious control measures are to use proper processes and containers. For home canning, Ball's Blue Book contains recommended procedures for all the low-acid canned foods listed in Table 17.9 except for corn and chicken mash, potato soup, and ripe and green olives. Chili peppers are not mentioned specifically but peppers are. Processes for all the acid foods listed in Table 17.9 may also be found in Ball's Blue Book.

All foods with pH less than 4.5 can be processed at 100°C (212°F). Foods with pH greater than 4.5 should be processed at 115.6°C (240°F) or higher. For home-canned foods, the process is carried out at either 100°C (212°C) or 115.6°C (240°F). No food should be canned if there is not a recommended time-temperature for processing it.

Media and Methods of Detection. Antitoxins for types A, B and E can be obtained directly from the Center for Disease Control in Atlanta, GA 30322. These antitoxins are necessary to protect mice used during testing to determine the types of botulinal toxin involved, as follows:

A portion of the sterile filtrate of the food homogenate (0.5 ml) is injected intraperitoneally into each of 2 mice. This food has been treated with trypsin to make the toxin more potent. Two other mice are injected intraperitoneally with 0.5 ml of undiluted, heated filtrate (100°C for 10 min). The mice are observed for symptoms of botulism within 72 hr. Deaths occurring with either the trypsinized or nontrypsinized, unheated filtrate and no deaths with the heated filtrate are considered to be presumptive evidence of toxin (FDA 1976).

To determine the type of toxin, mice are protected with antitoxin A, B or E. Unprotected mice (no antitoxin) are also injected with toxin. The mice are observed for 72 hr and all deaths are noted. If the unprotected mice die and the group of mice protected with either A, B or E antitoxin live, the presence of that particular type toxin is indicated.

Portions of the suspected food are added to freshly boiled and cooled chopped liver broth and to trypticase-peptone-glucose-yeast extract broth with sterile trypsin added (TPGYT). Both heated and unheated filtrates from these media are injected into 2 mice. If deaths occur with either or both the unheated filtrates and no deaths are noted with the heated filtrates, this constitutes presumptive evidence of toxin. Deaths are confirmed with passively protected mice (FDA 1976).

If the isolation of pure cultures is desired, 1 or 2 ml of the toxic culture are treated with ethyl alcohol (absolute) for 1 hr. A large loopful of the mixture is streaked onto anaerobic egg agar plates. Colonies are white to light yellow, flat and irregular, 1–2 mm in diameter, surrounded by a zone of yellow precipitate about 2 to 4 mm in diameter. The precipitated zone is irridescent to oblique light. This irridescent zone is characteristic of several other species of the genus *Clostridium*. For this reason, colonies are selected and grown in either TPGYT or freshly boiled chopped liver medium. The filtrates of these media are used to inject mice as previously described (FDA 1976). Also see Speck (1976) for more detailed procedures.

Investigation of a Botulism Outbreak. The data should be collected on a form from the Center for Disease Control (see Fig. 17.1) showing the specific foods eaten; and a relationship between the clinical symptoms and laboratory findings should be observed. In addition to the clinical syndrome, the epidemiological data should correlate or botulinal toxin should be found in a patient's blood or *C. botulinum* isolated from the food.

Selected Facts About *C. botulinum*

1.   Spores of *C. botulinum* are less sensitive to heat at pH values greater than 4.5 (Desrosier and Desrosier 1977). No higher *C. botulinum* resistance than 4 min at 120°C (248°F) and at a pH of 7 has been encountered (Perkins 1964).

2.   From 1899–1973, 292 deaths have occurred from toxin type A, 63 from B and 27 from E.

3.   A spore concentration of 60×10^{12} of A and B types can be reduced to 0 by heating them at 120°C (248°F) for 4 min at pH 7 (Perkins 1964).

4.   Type E spores have D values of 0.6 to 3.3 min at 80°C (176°F) (Schmidt 1964).

5.   If a low-acid canned food (with a pH above 4.5) is distributed at ambient temperatures equivalent to a process of 121.1°C (250°F) for 3 min to destroy *C. botulinum* spores fewer outbreaks will occur at home. The decline in cases since 1940 is probably due to improved canning methods in industry and at home (CDC 1970).

6.   Type A and B growth occurs at 10°C (50°F) and up to 48°C (118.4°F). Type E grows from 3.3° to 45°C (from 37.9° to 113°F) (Frazier 1967).

7.   There are seven types of *C. botulinum*. Types A and B are common types which affect man and are common in canned foods; Types C and D usually are associated with birds and mammals. Type G was isolated recently from soil in Argentina. Type E is associated with fish and fish products (CDC 1974).

8.   There is no growth of *C. botulinum* at pHs below 4.6 (National Canners Assoc. 1968).

9.   The limiting a_w values for growth according to Jay (1970) are:

Type A   0.94

Type B   0.94

Type E   0.97

10.   Vegetative growth of types A and B is inhibited at a_w 0.93–0.94. (Salt brine of 11.5% has a_w of 0.93.)

11.   Equivalent times (F values) to destroy type A spores of *C. botulinum* at different times are as follows (Esty and Meyer 1922):

(temp)	(min)
100°C (212°F)	360
105°C (221°F)	120
110°C (230°F)	36
115°C (239°F)	12
120°C (248°F)	4
121°C (250°F)	2.78

Schmidt (1964) obtained these D values of three strains at 80°C (176°F) for type E spores:

Strain	D Values at 80°C (176°F)
103	3.3
MINN	1.9
108	0.6

12. Type A and B spores are more heat-resistant than Type E (Perkins 1964).

13. California, Washington, Colorado, Oregon, and New Mexico account for over half of all the cases of reported outbreaks (CDC 1974). A geographic predilection for Alaska and the Great Lakes area is apparent for Type E.

14. Botulism can also occur from wounds (CDC 1974).

15. Botulinal toxins are extremely poisonous for man, and minute quantities acquired by ingestion, inhalation, or by absorption through the eye or a break in the skin can cause profound intoxication and death (CDC 1974).

16. Because of the low toxicity of some Type E, Type B, and Type F toxins, the toxicity is greatly enhanced by trypsin. Tests for toxicity should be performed with trypsinized, as well as untrypsinized materials (CDC 1974).

Bacillus cereus

Symptoms. Vomiting, diarrhea, nausea and pain are symptoms caused by *B. cereus*. The incubation time is about 15 hours.

Foods Involved. Data in Table 17.10 show that few cases of food poisoning are caused by *B. cereus*. The types of foods listed in Table 17.10 show that *B. cereus* may cause food poisoning in a diversified group of foods.

Prevention. *B. cereus* produces its toxin in the food. The bacterium is common in cereals, dust and soil. Its optimum growth temperature is 30°C (86°F). Minimum growth occurs at 10°–12°C (50°–53.6°C). It sporulates well in food. The spores are not very heat-resistant ($D_{100°C}$ 1 to 7.5 min). The method of control is to stop its growth in food by storing the food below its minimum growth temperature or above its maximum of 48°C (118.4°F)

TABLE 17.10.   FOOD INVOLVED IN FOODBORNE OUTBREAKS OF *BACILLUS CEREUS* POISONING

Year	Type of Food	Source
1969	Doughnuts	Restaurant
1969	Oysters	Home
1970	Gravy and dressing	Camp, restaurant, church
1971	Fried rice	Home

Source: CDC (1969, 1970, 1971).

and by using sanitary procedures in handling foods during storage and preparation.

Media and Methods of Detection. Media used to detect and identify *B. cereus* in foods are listed in Table 17.11. When Gram-positive rods, containing endospores, grow with a halo of dense precipitate around the colonies on phenol-egg yolk-polymyxin agar, this is a presumptive test for *B. cereus*. Other media, given in Table 17.11, are used in the identification of the bacterium. See Speck (1976) for additional instructions.

TABLE 17.11.   MEDIA USED IN DETERMINING *B. CEREUS* IN FOODS

Medium	Use	Reactions of *B. cereus*
Phenol red-egg yolk-polymyxin agar	Enumerate and isolate *B. cereus*	Colonies surrounded by a halo of dense precipitate with a violet red background
Phenol red carbohydrate broth	To determine acid and gas production from glucose, sucrose, glycerol and salicin	Acid from all; gas from none
Nitrate broth	To determine reduction of nitrate	Positive
Litmus milk	To determine peptonization	Positive
Nutrient gelatin	To determine liquefaction of gelatin	Positive
Starch agar	To determine hydrolysis of starch	Positive
V-P medium	To determine production of acetylmethylcarbinol	Positive

Source: FDA (1976).

Staphylococcus aureus

Symptoms. The symptoms for food poisoning caused by *S. aureus* are nausea, vomiting, diarrhea, abdominal pain and cramps, fever and chills. The onset of the symptoms is usually 6–8 hr but in some cases 24 hr.

Foods Involved. Data from the Center for Disease Control in Atlanta were compiled in Table 17.12. The one food that was the main vehicle of *Staphylococcus aureus* food poisoning during the years 1970–1974 was ham. For the most part, *Staphylococcus* food poisoning occurred with

TABLE 17.12.   FOOD INVOLVED IN FOODBORNE OUTBREAKS OF
STAPHYLOCOCCUS AUREUS POISONING

Year	Food	No. of Outbreaks	Locations
1970	Chicken salad	2	Home, school
	Turkey	3	Cafeteria, school
	Cake icing	1	Home
	Tacos	1	Home
	Chicken	2	Home, wedding
	Potato salad	1	Home
	Sardines	1	Home
	Turkey and potatoes	1	Restaurant
1971	Ham	4	Restaurant, party, home
	Turkey meat	1	Restaurant
	Raw pork dish	1	Luau
1972	Ham	13	[1]
	Pie	2	[1]
	Cake	1	[1]
	Eggs	1	[1]
	Fish	1	[1]
	Chicken	1	[1]
	Lima beans	1	[1]
	Roast beef	3	[1]
	Turkey salad	1	[1]
	Chopped liver	1	[1]
	Potato salad	1	[1]
1973	Ham	6	Home, school, hotel, restaurant
	Egg salad	2	Home, school
	French toast	1	Unknown
	Potato salad	1	Cafeteria
	Macaroni salad	1	School
	Lemon-filled jelly roll	2	Home
	Turkey	2	Home, church
1974	Ham	12	Restaurant, church, home, hospital
	Tuna salad	1	Home
	Spaghetti sauce	1	School
	Milk shake mix	1	Restaurant
	Milk	1	Home
	Pork	4	School, home, picnic

Source: CDC (1970, 1971, 1972, 1973, 1974).
[1] Location not given.

animal products but some plant foods were also involved (corn, lima beans). Food poisoning by this bacterium occurred in restaurants, cafeterias, homes, camps, schools, churches, bakeries, picnics, aboard aircraft, clubs, and even in hospitals.

Prevention. There are several ways to prevent staphylococci from entering food products. These are as follows (FDA 1967):

(a)   Employees should have good personal cleanliness.
(b)   Employees should have a health surveillance program.
(c)   There should be an effective sanitation program in the food processing plant and in the eating establishment.
(d)   Use proper temperature control. Keep the food below 4.4°C (40°F) or above 60°C (140°F).

(e) Destroy bacteria in raw products by approved methods, in milk and eggs by pasteurization.
(f) Have an effective control program; control insects, birds, rodents, dust, and moisture in the processing plant; wash, sanitize, handle, and store equipment properly.

Media and Methods of Detection. Some media commonly used for detecting and enumerating staphylococci are listed in Table 17.13. The Baird-Parker (BP) medium has gained popularity in the United States. The ingredients listed in Table 17.14 for the BP constitute the basal medium to which a 5% egg yolk emulsion and 0.3% of a 3.5% potassium tellurite are added.

On this substrate, black and shiny colonies with narrow white margins surrounded by clear zones extending into the opaque medium are counted as *S. aureus.* Colonies with this appearance are tested for coagulase production.

Vogel and Johnson agar, Baird-Parker medium and tellurite-polymyxin-egg yolk agar contain tellurite to aid in determining the staphylococci. Formulae for above are listed in Table 17.14. Salt concentration in the TPEYA and S110 acts as a selective agent for staphylococci. See Speck (1976) for additional procedures.

TABLE 17.13. MEDIA USED TO DETECT *STAPHYLOCOCCUS*

Medium	Reference
Vogel and Johnson Agar (VJA)	BBL (1968)
Baird-Parker Medium (BPA)	FDA (1976)[1]
Tellurite-polymyxin-egg yolk agar (TPEYA)	Lewis and Angellotti (1964)
Staphylococcus 110 (S110)	Chapman (1946)

[1] FDA Bacteriological Analytical Manual (1976).

Salmonella

Symptoms. Salmonellae cause gastrointestinal disorders. Symptoms are fever, cramps, diarrhea and sometimes vomiting.

Foods Involved. Foods listed in Table 17.15 served as vectors in food infections due to salmonellae during the years of 1969, 1970, 1971, 1973 and 1974. Turkey, beef, ham and ice cream were leading vectors of salmonellae. Veal, pork, beef, poultry and fish are usually associated with salmonellae infections when food is the vehicle.

Prevention. Some of the means listed for staphylococci are also relevant for prevention of contamination with *Salmonella* sp. The preventive methods are as follows (FDA 1969):

(a) Promote personal cleanliness.

TABLE 17.14.   FORMULAE OF MEDIA USED TO DETECT *STAPHYLOCOCCUS*
(Ingredients in g/liter)

VJA[1]		BPA[2]		TPEYA[3]		S110[4]	
Trypticase-peptone	10	Tryptone	10	Tryptone	10	Tryptone	10
Yeast extract	5	Beef extract	5	Yeast extract	5	Yeast extract	2.5
Mannitol	10	Yeast extract	10	Mannitol	5	Lactose	2
Dipotassium phosphate	5	Sodium pyruvate	5	Sodium chloride	20	Mannitol	10
Potassium tellurite of 1% solution	20 (ml)	Glycine	12	Lithium chloride	2	Sodium chloride	75
Lithium chloride	5	Lithium chloride·6H$_2$O	5	Agar	18	Dipotassium hydrogen phosphate	5
Glycine	10	Agar	20	Polymyxin B sulphate of 1% (ml) solution	0.4	Gelatin	30
Agar	16	For enrichment add Bacto EY Tellurite		Egg yolk:saline emulsion (ml)	100	Agar	15
Phenol red	25 mg			Potassium tellurite of 1% solution (ml)	10		

[1] Source: BBL (1968).
[2] Source: FDA (1976).
[3] Source: Lewis and Angelotti (1964).
[4] Source: Chapman (1946).

TABLE 17.15.   FOODS ASSOCIATED WITH CONFIRMED CASES OF FOOD
INFECTION CAUSED BY *SALMONELLA*

Year	Food	No. of Times Food Vehicle	Locations
1969	Custard doughnuts	1	Bakery, home
	Custard cake	1	Bakery, home
	Meatballs	1	Delicatessen, home
	Roast beef	3	Restaurant, caterer, home
	Chicken	2	Home
	Ice cream	1	Home
	Caesar salad	1	Banquet
	Cream pie	1	School
	Turkey	8	Hospital, caterer, home, banquet
	Pound cake	1	Home
	Chicken salad	1	Home
	Chicken and/or eggs	1	Picnic
1970	Turkey	6	Home, caterer, camp, restaurant
	Chicken	2	Restaurant, nursing home
	Egg nog	1	Home
	Eggs	1	Home
	Ham	5	School, restaurant, home
	Pork	1	Restaurant
	Ice cream	4	Home, school
	Tacos	1	Restaurant
	Hamburger	1	Restaurant
	Roast beef	1	Restaurant
	Raw beef	1	Cafeteria
1971	Roast pork	1	Home
	Chef, shrimp and tossed salads	1	Restaurant
	Deviled eggs, ham dip	1	Country club
	Chicken salad	1	Home
	Turkey and rice stuffing	1	Home
	Roast beef	1	Home
	Lemon meringue pie	1	Home
	Pork spare ribs	1	Restaurant
	Chicken	1	Restaurant
	Turkey	1	Home
	Beef stew	1	Restaurant
	Chicken spread	1	Home
	Fruit salad	1	Picnic
1972	Raw beef	3	1
	Fat back	1	1
	Custard	1	1
	Ice cream	4	1
	Turkey	1	1
	Ham	1	1
	Beef	3	1
	Coconut cream pie	1	1
	Gravy	1	1
	Deviled eggs	1	1
	Cole slaw	1	1
	Bread dressing	1	1
	Boiled salmon	1	1
	Chicken	1	1

TABLE 17.15. (*Continued*)

Year	Food	No. of Times Food Vehicle	Locations
	Pork	1	[1]
	Head cheese	1	[1]
	Potato salad	1	[1]
1973	Beef	6	Restaurant, home, ship
	Ice cream	4	Home, camp
	Eggs	1	Hospital
	Turkey	3	Church, school, fraternity house
	Fish	1	Church
1974	Ice cream	6	Home, church
	Potato salad	1	Barbecue
	Barbecue pork	1	Home
	Barbecue sandwich	1	Restaurant
	Chicken salad	2	Parish, wedding reception

Source: CDC (1969, 1970, 1971, 1972, 1973, 1974).
[1] Location not given.

(b) Adopt an employee health surveillance program.
(c) All employees should be trained to prepare, handle and store foods in a sanitary manner.
(d) Adopt good manufacturing practices.
(e) Maintain rigid control on all incoming ingredients and reject all unfit raw materials.
(f) Destroy bacteria in raw products by approved methods.
(g) Maintain proper storage temperatures.
(h) Rotate raw and finished products and destroy spoiled foods.
(i) Eliminate insects, birds, and rodents.
(j) Control dust in plant.
(k) Assure clean air intake systems.

Media and Methods of Detection. Media used to detect salmonellae are named in Table 17.16. The ingredients in each medium are listed in Table 17.17. If BGA, SS and BSA are used, typical colonies are picked and used with TSI. Isolates that conform to *Salmonella* are then studied by biochemical and serological methods to determine the specific type present in the food.
See Speck (1976) for additional information.

Shigella

Symptoms. Shigellosis is characterized by sudden appearance of abdominal pain, cramps, diarrhea, fever, and vomiting; blood, pus and mucus may be found in the stools of about $\frac{1}{3}$ of the patients.
Foods Involved. *Shigella* was found in foods listed in Table 17.18 during 1970 to 1974. No specific food was the dominant vector of shigellosis.

TABLE 17.16.   MEDIA USED TO DETECT *SALMONELLA*

Medium	Use
Lactose broth (LB)	Enrichment
Selenite-cystine broth (SCB)	Enrichment
Tetrathionate broth (TB)	Enrichment
Brilliant green agar (BGA)	*Salmonella* colonies pink to fuchsia, translucent to opaque with surrounding medium pink to red
Salmonella-Shigella agar (SSA)	Uncolored to pale pink, opaque, transparent. Some strains produce black-centered colonies
Bismuth-sulfite agar (BSA)	Colonies brown, black, sometimes with metallic sheen. Some strains produce green colonies
Triple sugar-iron agar (TSI)	Slants are alkaline (red) and butts are acid (yellow) with or without H_2S formation (blackening of agar)

Source: FDA (1976).

Prevention. The bacteria are transmitted directly by active cases, by convalescent patients and by healthy carriers. Control procedures are as follows (FDA 1969):

(a)   Promote personal cleanliness among employees.

(b)   Provide adequate toilet facilities for employees and require handwashing after toilet use.

(c)   Do not allow persons sick with abdominal cramps and diarrhea to prepare or handle food.

(d)   Dispose of sewage properly.

(e)   Institute fly control and prevent fly breeding by keeping the plant and surroundings clean.

(f)   Maintain proper cold storage temperatures of perishable foods.

(g)   Provide periodic employee instruction in the sanitary preparation, handling, and storage of foods.

Media and Methods of Detection. Media for detecting *Shigella* in foods are listed in Table 17.19. Gram-negative broth and selenite-cystine broth are used for enrichment. Xylose-lysine-desoxycholate agar, desoxycholate-citrate agar, and eosin-methylene blue agar are used as selective media. The reactions of *Shigella* on these media are also given in Table 17.19. The reader is referred to Table 17.20 for formulae of these substrates.

See Speck (1976) for additional details.

TABLE 17.17.   FORMULAE OF MEDIA USED TO DETECT *SALMONELLA*
(Ingredients in g/liter)

LB¹		SCB²		TB²		BGA³		SSA¹		BSA¹	
Beef extract	3	Polypeptone peptone	5	Polypeptone peptone	5	Proteose peptone	10	Beef extract	5	Polypeptone	10
Peptone	5	Lactose	4	Bile salts	1	Yeast extract	3	Polypeptone peptone	5	Beef extract	5
Lactose	5	Sodium phosphate	10	Calcium carbonate	10	Sodium chloride	5	Lactose	10	Dextrose	5
DW (liter)	1	Sodium acid selenite	4	Sodium thiosulfate	30	Lactose	10	Bile salt mixture	8.5	Disodium phosphate (anhydrous)	4
		L-cystine	0.01	DW (liter)	1	Saccharose	10	Sodium citrate	8.5	Ferrous sulfate (anhydrous)	0.3
		DW (liter)	1	Add 20 ml of iodine solution to basal medium (iodine sol'n. contains iodine, 6 g; potassium iodide, 5 g; DW, 20 ml)		Brilliant green	0.0125	Sodium thiosulfate	8.5	Bismuth sulfite indicator	8
						Phenol red	0.08	Ferric citrate	1	Brilliant green	0.025
						Agar	20	Agar	13.5	Agar	20
						DW (liter)	1	Brilliant green	0.00033 (0.33 mg)	DW (liter)	1
								Neutral red	0.025 (25 mg)		
								DW (liter)	1		

¹ Source: FDA (1976).
² Source: BBL (1968).
³ Source: Difco Laboratories (1953).

TABLE 17.18. FOODS ASSOCIATED WITH CONFIRMED CASES OF INFECTION CAUSED BY *SHIGELLA* IN FOOD

Year	Food	Location
1970	Pudding	School
1971	Chicken spread	Home
	Fruit salad	Picnic
1972	Strawberries	[1]
1973	Chopped turkey	School
	Fish salad	Restaurant
	Shrimp salad	Hospital
	Rice balls	Luau
	Tuna fish salad	School
1974	Potato salad	Home

Source: CDC (1970, 1971, 1972, 1973, 1974).
[1] Location not given.

TABLE 17.19. MEDIA USED TO DETECT *SHIGELLA*

Medium	Use
Gram-negative broth (GNB)	For enrichment of *Shigella*
Selenite-cystine broth (SCB)	For enrichment of *Shigella*
Xylose-lysine-desoxycholate agar (XLD)	*Shigella* colonies appear to be rose-colored, surrounded by a rosy halo when viewed by transmitted light
Desoxycholate-citrate agar (DCA)	Colonies gray or slate gray; entire colonies glistening by reflected light
Eosin-methylene blue agar (EMB)	As on DCA
Triple sugar-iron agar (TSI)	*Shigella* causes alkaline (red) slant and acid (yellow) butt without gas and H_2S production

Source: FDA (1976).

Vibrio parahaemolyticus

Symptoms. The symptoms for food poisoning caused by *V. parahaemolyticus* are epigastric pain, nausea, vomiting, and diarrhea with occasional blood and mucus in the stools. A fever of 0.6°–1.1°C (1°–2°F) is experienced in 60–70% of the cases. The onset of the symptoms is 15–17 hr and they last for 1–2 days.

Foods Involved. This bacterium is found in the seas. As shown in Table 17.21, it is associated mostly with seafoods.

Prevention. Those recommendations pertaining to sanitation and sanitary handling of foods also apply here. Illness caused by this bacterium can best be avoided by eating only cooked seafoods.

Media and Methods of Detection. Because of the microorganism's relation to salt, all media must contain 2–3% NaCl. A 3% NaCl dilution blank is used in isolating the bacterium from foods.

The food may be enriched in glucose salt teepol broth (GSTB) overnight.

TABLE 17.20. FORMULAE OF MEDIA USED TO DETECT *SHIGELLA*
(Ingredients in g/liter)

GNB[1]		SCB[2]		XLD[2]		DCA[3]	
Polypeptone	20	Polypeptone	5	Yeast extract	3	Meat infusion	330
Dextrose	1	Lactose	4	L-lysine	5	Peptone	10
D-mannitol	2	Sodium acid selenite	4	Xylose	3.50	Lactose	10
Sodium citrate	5	Sodium phosphate	10	Lactose	7.5	Sodium citrate	20
Sodium desoxycholate	0.5	L-cystine	0.01	Sucrose	7.5	Sodium desoxycholate	5
Dipotassium phosphate	4	DW (liter)	1	Sodium desoxycholate	2.5	Ferric ammonium citrate	2
Monopotassium phosphate	1.5			Sodium thiosulfate	6.8	Agar	13.5
Sodium chloride	5			Sodium chloride	5	Neutral red	0.02
DW (liter)	1			Agar	13.5	DW (liter)	1
				Phenol red	0.08		
				Ferric ammonium citrate	0.8		
				DW (liter)	1		

EMB[2]		TSI[1] (Ingredients No. 1)		TSI[1] (Ingredients No. 2)	
Peptone	10	Polypeptone	20	Beef extract	3
Lactose	10	Sodium chloride	5	Yeast extract	3
Dipotassium phosphate	2	Lactose	10	Peptone	15
Agar	15	Sucrose	1	Proteose peptone	5
Eosin Y	0.4	Glucose	1	Glucose	1
Methylene blue	0.065	Ferric ammonium sulfate	0.2	Lactose	10
DW (liter)	1	Sodium thiosulfate	0.2	Sucrose	10
		Phenol red	0.025	Ferrous sulfate	0.2
		Agar	13	Sodium chloride	5
				Sodium thiosulfate	0.3
				Phenol red	0.024
				Agar	12

[1] Source: FDA (1976).
[2] Source: BBL (1968).
[3] Difco Laboratories (1953).

TABLE 17.21.   FOODS ASSOCIATED WITH CONFIRMED CASES OF *VIBRIO PARAHAEMOLYTICUS*

Year	Food	Location
1971	Steamed crabs	Picnic
	Crab salad	Hospital
1972	Crab	1
	Shrimp	1
	Lobster salad	1
1973	Conch meat	Home

Source: CDC (1971, 1972, 1973).
[1] Location not given.

The bacteria are then streaked on thiosulfate-citrate-bile salts-sucrose agar. Formulae for these media are given in Table 17.22.

Colonies which are 2–3 mm in diameter with green or blue centers are presumptive evidence of *V. parahaemolyticus.* Biochemical and serological evidence is needed for a more thorough identification.

Another medium, the modified Twedt (MT), consisting of 2% peptone, 0.2% yeast extract, 1% corn starch, 7% NaCl and 1.5% agar (pH 8.0), is used to select and differentiate the organism. On this substrate, *V. parahaemolyticus* colonies appear white to creamy, circular, smooth and amylase-positive.

See Speck (1976) for additional information.

TABLE 17.22.   MEDIA USED TO DETECT *V. PARAHAEMOLYTICUS*
(Ingredients in g/liter)

Glucose-Salt-Teepol Broth (GSTB)		Thiosulfate-Citrate-Bile Salts-Sucrose Agar (TCBS)	
Beef extract	3	Yeast extract	5
Peptone	10	Peptone	10
NaCl	30	Sucrose	20
Glucose	5	Sodium thiosulfate	10
Methyl violet	0.002	Sodium citrate	10
Teepol[1] (ml)	4	Sodium cholate	3
DW (liter)	1	Oxgall	5
(Single strength)		Sodium chloride	10
		Ferric citrate	1
		Bromthymol blue	0.04
		Thymol blue	0.04
		Agar	15
		DW (liter)	1

Source: FDA (1976).
[1] Available through Shell Chemical Co., 110 West 51st, New York, NY 10020.

Streptococcus

Symptoms. *Streptococcus* causes an infection. The symptoms are nausea, vomiting and diarrhea.

Foods Involved. Foods associated with food infections during 1969, 1972, 1973 and 1974 are given in Table 17.23. Streptococcal infections are rare when compared to occurrences of staphylococcal and salmonellae food poisoning and food infections.

Prevention. All the recommendations under the prevention of other food poisoning and food infectious bacteria should be followed.

Media and Methods of Detection. Fecal streptococci are grown on KF streptococcus agar. This medium consists of 10 g proteose peptone, 10 g yeast extract, 5 g sodium chloride, 10 g sodium glycerophosphate, 20 g maltose, 1 g lactose, 0.4 g sodium azide, 0.015 g bromcresol purple and 20 g agar. This medium is heated to boiling, then autoclaved for 10 min at 121°C (249.8°F). The medium is cooled to 60°C (140°F) and 1 ml of 1% solution Triphenyltetrazolium chloride is added per 100 ml of media. The final pH is 7.2.

Dark red colonies or those having a red or pink center are considered to be streptococci.

See Speck (1976).

TABLE 17.23. FOODS ASSOCIATED WITH CONFIRMED CASES OF *STREPTOCOCCAL* FOOD INFECTIONS

Year	Food	Location
1969	Creamed shrimp	Restaurant
1972	Cod fish	[1]
	Frankfurters	[1]
1973	Potato salad	Picnic
1974	Egg salad	County jail
	Lomi salmon	Home
	Barbecue	Church

Source: CDC (1969, 1972, 1973, 1974).
[1] Location not given.

MYCOTOXINS

Filamentous fungi or molds can grow on various foodstuffs and produce poisons called mycotoxins. Examples of mycotoxins and their toxicity are listed in Table 17.24. Some fungi which are the source of these mycotoxins are found in Table 17.25.

TABLE 17.24.   EXAMPLES OF MYCOTOXINS AND THEIR TOXICITY

Mycotoxin	Toxicity	References
Aflatoxins	Carcinogenic in test animals. Toxins may be involved in the etiology of human liver cancer. B_1 is most toxic. Cows eat B_1 and excrete M_1	AOAC (1975)
Sterigmatocystin	Substance is toxic and carcinogenic	AOAC (1975)
Penicillic acid	Mammalian toxicant carcinogenic to rats and mice	Thorpe and Johnson (1974)
Ochratoxin	Type A carcinogenic to trout. Causes kidney and liver damage in some animals	AOAC (1975)
Patulin	Compound is an antibiotic, with carcinogenic and mutagenic properties	Ware et al. (1974)
T-2 toxin	Toxin produces skin irritations	Chung et al. (1974)
Zearalenone	An estrogenic, secondary fungal metabolite. Causes hyperestrogenism when ingested by swine	Mirocha et al. (1974)
Citrinin	Causes lesions of the kidney in rats, rabbits, guinea pigs and swine	Hald and Krogh (1973)

TABLE 17.25.   EXAMPLES OF FUNGI WHICH PRODUCE SPECIFIC MYCOTOXINS

Mycotoxin	Fungi-Producing Mycotoxin	References
Aflatoxin	Aspergillus flavus A. parasiticus Penicillium puberulum	Frazier (1967) Frazier (1967) Frazier (1967)
Sterigmatocystin	Aspergillus versicolor A. nidulans	Stack and Rodricks (1973)
Penicillic acid	Penicillium puberulum	Thorpe and Johnson (1974)
Ochratoxin	Aspergillus ochraceus A. flavus	Levi et al. (1974) Levi et al. (1974)
Patulin	Penicillium expansum	Lovett et al. (1975)
T-2 toxin	Fusarium sp.	Lovett et al. (1975)
Zearalenone	Fusarium roseum	Eppley et al. (1974)
Citrinin	Penicillium citrinum Aspergillus niveus A. terreus	Raper and Thom (1949) Thom and Raper (1945) Raper and Thom (1949)

Aflatoxins

There are several types of aflatoxins which vary only slightly in chemical structure. The types are B_1, B_2, G_1, G_2, M_1 and M_2. Some strains produce more than one aflatoxin. For example, ATCC strain 15517 of *Aspergillus parasiticus* produces B_1, G_1, B_2, G_2 (ATCC 1974). Types B_1 and G_1 are produced by *Aspergillus flavus* ATCC 15546 and B_2 by *A. flavus* ATCC 24109. Some strains of *Aspergillus flavus* do not yield any aflatoxins. Chah (1974) studied the effects of fungus-fermented soybeans on broiler growth (chickens) and life cycle performance of *Coturnix* quail. He found that some strains of *Aspergillus oryzae, A. flavus, A. sydowi, A. clavatus, A. candidus, A. restrictus, A. aurecomus, A. elegans, A. fischeri, A. tamarii* and *A. janus* produced no mycotoxins in their fermentations.

Foods Involved. Wheat, rice, peanuts and peanut products, corn and corn products, cottonseed and cottonseed meal have all been adequate substrates for the growth of producers of mycotoxins. The appearance of colonies of *A. flavus* is illustrated in Fig. 17.3.

Environmental Factors Affecting Aflatoxin Production.

(a) Short periods of high temperature cause depression of growth of the fungus and accumulation of toxin.

(b) Optimum temperature for maximum yields of both B_1 and G_1 on rice was 28°C (82.4°F). At 37°C (98.6°F) growth was good but production was low (300–700 ppb).

(c) In a casein substrate, *A. flavus* and *A. parasiticus* were tested for aflatoxin over a pH range of 1.0 to 11.0. After 10 days, the pH of nearly all samples approached neutrality. Aflatoxins B_1 and G_1 were produced at all pH values where growth occurred. The highest levels, generally, were present after 21 days in samples which were initially the extreme ends of the pH scale.

Studies Involving Man. Bourgeois (1972) had the opportunity to study children in Thailand who ate foods containing aflatoxin B_1. The children died of Reye's syndrome. Assays of the food demonstrated that

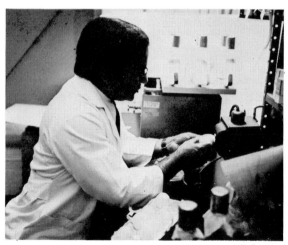

Courtesy of University of Missouri-Columbia

FIG. 17.2. FERTILE EGGS ARE USED TO TEST FOR THE PRESENCE OF MYCOTOXINS

Courtesy of Food and Drug Administration

FIG. 17.3. THESE WALNUT MEATS SHOW A HEAVY CONTAMINATION BY *ASPERGILLUS FLAVUS,* THE MOLD SPECIES WHICH PRODUCES AFLATOXINS, GROWN UNDER IDEAL CONDITIONS IN THE LABORATORY

Aflatoxin is one of a number of toxins, called mycotoxins, produced by various molds.

about 5000 ppm of aflatoxin B_1 and high concentrations of four other unidentified toxic substances were present. Four toxigenic fungi were isolated from the food: *Aspergillus flavus,* two strains of *Aspergillus* and a *Penicillium.*

Assays for aflatoxin in the brain, liver, kidney, intestinal content and stool specimens were made on 22 autopsied cases. Concentrations of aflatoxin ranging from a trace to 127 μg per kg were found in the intestinal contents and stool specimens in 96% of the cases whereas the tissue levels ranged from a trace to 93 μg per kg.

Young cynomolgous monkeys were fed aflatoxin B_1 produced by a strain of *Aspergillus* isolated from a rice sample from which the children had eaten. Upon autopsy, myocardial, renal, cerebral and hepatic lesions virtually identical to Reye's syndrome were demonstrated (Bourgeois 1972).

Bourgeois (1972) concluded from his study of Thai children and from studies with short-tailed monkeys that there were similar combinations of clinical, laboratory and pathological findings. These similarities between these two conditions plus epidemiological data strongly suggest that there was an etiologic role of aflatoxins in encephalopathy and fatty degeneration of the viscera of the Thai children (Bourgeois 1972). More research in the future will explain the specific effects of aflatoxins on human beings.

Methods of Analysis. The *Official Methods of Analysis of the Association of Official Analytical Chemists, 12th Edition* (AOAC 1975) recommends thin-layer chromatography and chicken embryo bioassay for analyzing aflatoxin in peanuts and peanut products, coconut, copra, copra meal, cottonseed products, dairy products, green coffee, pistachio nuts, and soybeans. In Fig. 17.2 a scientist demonstrates the technique used to detect mycotoxins.

Ochratoxin

As indicated in Table 17.25, ochratoxin is produced by *Aspergillus ochraceus* and *A. flavus.*

Foods Involved. The Official Methods of AOAC (1975) gives a method for assaying barley for ochratoxins. *A. ochraceus* has been isolated from sorghum grain. Corn, peanuts and Brazil nuts will support growth and toxin production.

Types of Ochratoxin. Two types of ochratoxin, A and B, have been isolated. These two compounds differ only in a chlorine molecule. Strain ATCC 18412 of *A. ochraceus* produces type A ochratoxin.

Patulin

Patulin is produced by strain NRRL 976 of *Penicillium expansum* which causes rot in apples. The toxin has been found in apple juice in which the fungus has grown. However, the AOAC (1975) does not designate any specific food for analysis by thin-layer chromatography.

Sterigmatocystin

Sterigmatocystin is produced by *Aspergillus versicolor* and *A. nidulans* (Table 17.25). This mycotoxin has been found in cornmeal and probably other agricultural commodities. It is assayed by preliminary thin-layer chromatography and by column chromatography followed by thin-layer chromatography.

Ways of Controlling Mycotoxins

The FDA (1966) recommendations of ways of controlling mycotoxins are as follows:

1. Prevention of mold growth by proper drying and storage of crops.
2. Removal of mold-damaged material before storage or processing.
3. Proper feed mill sanitation with adequate moisture and humidity control of stored feedstuffs.
4. Use of antifungal substances which must *meet food additive* requirements to be acceptable.
5. Feeding moldy material cautiously to a few animals which are watched closely for any signs of illness. Disease-producing lots will have to be destroyed in many instances.
6. If necessary, having chemical and biological assay made routinely on some materials, such as corn or peanut and cottonseed meals, to detect the presence of certain known mycotoxins.

GLOSSARY

Antitoxin—An antibody formed in response to and capable of neutralizing a poison of biological origin.

Bile salts—Salts contained in bile. Bile is an alkaline brownish-yellow or greenish-yellow liquid that is secreted by the liver.

Brilliant green—Brilliant green, also used in dyeing silk, wool, jute and cotton, is a topical antiseptic in infected wounds. It is also a medicine for canine dysentery and a staining constituent of bacteriological media.

Conch meat—Any of various tropical marine gastropod mollusks of the genus *Strombus* and other genera having large, often brightly colored, spiral shells and edible flesh.

Eosin—A red dye.

Hemolysin—An agent or substance that initiates lysis of red blood cells, thereby liberating hemoglobin.

Indole—Metabolite of tryptophan.

Kanamycin—An antibiotic produced by *Streptomyces kanamyceticus.*

Methylene blue—Dark green, odorless crystals used as a stain in bacteriology; also as an oxidation-reduction indicator; an antiseptic.

Oxgall—Oxgall is dehydrated fresh bile. It is used to restrict growth of fungi and is selective for intestinal bacteria.

Phenol red—A pH indicator.

Polymyxin B—Antibiotic produced by *Bacillus polymyxa.* It is effective against Gram-negative bacteria.

Selenite—Gypsum in the form of colorless clear crystals. Gypsum is $CaSO_2 \cdot 2H_2O$.

Sodium cholate—Sodium salt of cholic acid. This compound is obtained from ox bile extract.

Sodium desoxycholate—Makes media selective for coliform bacteria.

Sodium tellurite—Sodium or potassium salt of telluric acid. Staphylococci growing on media containing this salt produce black colonies.

Sodium thioglycollate—Lowers the Eh of media.

Stormy fermentation—A very vigorous fermentation with large quantities of gas.

Sulfadiazine—Antimicrobial agent against streptococci, staphylococci, salmonellae and coccida.

REVIEW QUESTIONS

1.   What are the symptoms caused by *Clostridium perfringens*? How soon is the onset of symptoms?
2.   What foods are usually involved in outbreaks of food poisoning caused by *C. perfringens*?
3.   Who keeps records on the outbreaks of food poisoning?
4.   How can you prevent food poisoning by *C. perfringens*?
5.   Describe how to isolate and detect *C. perfringens*.
6.   How is the toxin tested?
7.   How can one estimate the previous growth of *C. perfringens*?
8.   What kind of epidemiological data are collected?
9.   What are the symptoms caused by *Clostridium botulinum*?
10.   What foods are usually involved?
11.   What method of food preservation causes the highest incidence of botulism?
12.   How does one prevent botulism?
13.   How does one detect *C. botulinum* in food?
14.   How does one determine preformed toxin in food?
15.   Explain how to isolate *C. botulinum*.
16.   Explain how to detect toxin in blood serum.
17.   What animals are used to test botulinal toxin?
18.   What type of *C. botulinum* has the most heat-resistant bacterial spores?
19.   What determines if a food should be heated at 100°C (212°F) or 115.5°C (240°F) during canning?

20. At what pH are spores of *C. botulinum* most heat-resistant?
21. What effect do curing ingredients have on heated spores (given mild heat treatment)?
22. For what length of time should low-acid foods be heated at 121.1°C (250°F) to kill the spores of *C. botulinum*?
23. At how low of a temperature do types A and B grow? Type E?
24. What pH limits growth of type A, B, and E?
25. What brine strength limits the growth of *C. botulinum*?
26. How heat-resistant is type E?
27. What effect does HCl (concentration in stomach) have on botulinal toxin?
28. What effect can other bacteria have on botulinal toxin?
29. What are the symptoms caused by *Bacillus cereus* food poisoning?
30. What foods are involved in food poisoning by *B. cereus*?
31. How can food poisoning by *B. cereus* be prevented?
32. What media are used to isolate *B. cereus*?
33. How do you identify *B. cereus*?
34. What are mycotoxins?
35. How many kinds of aflatoxins are there?
36. What is type M aflatoxin associated with?
37. What fungi produce aflatoxins?
38. Do all strains of *Aspergillus flavus* produce aflatoxins?
39. List some mycotoxins and the fungi which produce them.
40. What adverse effects do mycotoxins have on animals?
41. What foods are involved in aflatoxins, ochratoxins, patulin and sterigmatocystin?
42. How can mycotoxins be controlled?

REFERENCES

ATCC. 1974. Catalogue of Strains, 11th Edition. American Type Culture Collection, Rockville, Maryland.

AOAC. 1975. Official Methods of Analysis. Association of Official Analytical Chemists, Washington, D.C.

BBL. 1968. BBL Manual of Products and Laboratory Procedures, 5th Edition. BBL, Division of Becton, Dickinson and Co., Cockeysville, Maryland.

BOURGEOIS, C. H. 1972. Encephalopathy and fatty degeneration of the viscera as a possible response to aflatoxins. Proc. Symp. Mycotoxins and Mycotoxicoses. University of Missouri-Columbia.

BREED, R. S. and MURRAY, E. G. D. 1957. Bergey's Manual of Determinative Bacteriology, 7th Edition. Williams & Wilkins Co., Baltimore.

BUCHANAN, R. E. and GIBBONS, N. E. 1974. Bergey's Manual of Determinative Bacteriology, 8th Edition. Williams & Wilkins Co., Baltimore.

CDC. 1969, 1970, 1971, 1972, 1973, and 1974. Foodborne Outbreaks. USDHEW
Center for Disease Control, Atlanta, Georgia.

CDC. 1974. Botulism in the United States, 1899–1973. Handbook for Epide-
miologists, Clinicians, and Laboratory Workers. USDHEW Center for Disease
Control, Atlanta, Georgia.

CHAH, C. C. 1974. Effects of fungus-fermented soybeans on broiler growth and
life cycle performance of Coturnix quail. Ph.D. Thesis, S. Dakota State University.

CHAPMAN, G. H. 1946. A single culture medium for selective isolation of
plasma-coagulating Staphylococci and for improved testing of chromogenesis,
plasma coagulation, and the stone reaction. J. Bacteriol. *51*, 409–410.

CHUNG, C. W., TRUCKSESS, M. W., GILES, A. L., JR. and FRIEDMAN, L.
1974. Rabbit skin test for estimation of T-2 toxin and other skin-irritating toxins
in contaminated corn. J. AOAC *57*, 1121–1127.

DESROSIER, N. W. and DESROSIER, J. N. 1977. The Technology of Food
Preservation, 4th Edition. AVI Publishing Co., Westport, Conn.

DIFCO LABORATORIES. 1953. Difco Manual of Dehydrated Culture Media
and Reagents for Microbiological and Clinical Laboratory Procedures, 9th Edition. Difco Laboratory Procedures, 9th Edition. Difco Laboratories, Detroit.

EPPLEY, R. M., STOLOFF, L., TRUCKSESS, M. W. and CHUNG, C. W. 1974.
Survey of corn for *Fusarium* toxins. J. AOAC *57*, 632–635.

ESTY, J. R. and MEYER, K. F. 1922. The heat resistance of the spores of *C.
botulinum* and allied anaerobes. Part II. J. Infect. Diseases *31*, 650–663.

FDA. 1966. Fact sheet on mycotoxins. Food and Drug Administration, US-
DHEW, Washington, D.C.

FDA. 1967. Fact sheet on Staphylococci in food. Food and Drug Administration,
USDHEW, Washington, D.C.

FDA. 1969. Fact sheet on *Shigella* in food. Food and Drug Administration,
USDHEW, Washington, D.C.

FDA. 1969. Fact sheet on *Salmonella* in foods. Food and Drug Administration,
USDHEW, Washington, D.C.

FDA. 1976. Bacteriological Analytical Manual for Foods (BAM), 4th Edition.
Association of Official Analytical Chemists, Washington, D.C.

FRAZIER, W. C. 1967. Food Microbiology, 2nd Edition. McGraw-Hill Book
Co., New York.

HALD, B. and KROGH, P. 1973. Analysis and chemical confirmation of citrinin
in barley. J. AOAC *56*, 1440–1443.

JAY, J. M. 1970. Modern Food Microbiology. Van Nostrand Reinhold Co., New
York.

LEVI, C. P., TRENK, H. L. and MOHR, H. K. 1974. Study of the occurrence of
ochratoxin A in green coffee beans. J. AOAC *57*, 866–870.

LEWIS, K. H. and ANGELOTTI, R. 1964. Examination of foods for entero-
pathogenic and indicator bacteria. *In* Review of Methodology and Manual of
Selected Procedures. USDHEW, U.S. Government Printing Office, Washington,
D.C.

LOVETT, J., THOMPSON, R. G., JR. and BOUTIN, B. K. 1975. Trimming as
a means of removing patulin from fungus-rotted apples. J. AOAC *58*, 909–
911.

MIROCHA, C. J., SCHAUERHAMER, B. and PATHRE, S. V. 1974. Isolation, detection and quantitation of zearalenone. J. AOAC *57*, 1104–1110.

NATIONAL CANNERS ASSOCIATION. 1968. Laboratory Manual for Food Canners and Processors, Vol. 1 and 2. AVI Publishing Co., Westport, Conn.

PERKINS, W. E. 1964. Prevention of botulism by thermal processing. *In* Botulism, Proceedings of a Symposium. USDHEW, Cincinnati.

RAPER, K. B. and THOM, C. 1949. A Manual of the Penicillia. Williams & Wilkins Co., Baltimore.

SCHMIDT, C. W. 1964. Spores of *C. botulinum*: formation, resistance, germination. *In* Botulism, Proceedings of a Symposium. USDHEW, Cincinnati.

SMITH, L. D. S. 1955. Introduction to the Pathogenic Anaerobes. University of Chicago Press, Chicago.

SMITH, L. D. S. 1977. Botulism: The Organisms, Its Toxins, the Disease. Charles C. Thomas, Publisher, Springfield, Illinois.

SPECK, M. L. (Editor). 1976. Compendium of Methods for the Microbiological Examination of Foods. American Public Health Association.

STACK, M. and RODRICKS, J. V. 1973. Collaborative study of the quantitative determination and chemical confirmation of sterigmatocystin in grains. J. AOAC *54*, 1123–1125.

THOM, C. and RAPER, K. B. 1945. A Manual of the Aspergilli. Williams & Wilkins Co., Baltimore.

THORPE, C. W. and JOHNSON, R. L. 1974. Analysis of penicillic acid by gas-liquid chromatography. J. AOAC *57*, 861–865.

WARE, G. M., THORPE, C. W., and POHLAND, A. E. 1974. Liquid chromatographic method for determination of patulin in apple juice. J. AOAC *57*, 1111–1113.

18

Methods of Study and Review for Food Microbiology Examinations

METHODS OF STUDY

Note Taking

The taking or writing of notes is a skill that one develops with time. It is important that the student take the notes in an organized manner in outline form. Of course, if the lecturer is disorganized, the task is more difficult. If a person is taking notes on material from books or journals and proceeding at his own pace, it is easier to keep the material in its proper order. The note taker must be selective and not try to take every word. He should note the relationship and the principle and the example which illustrates them.

Rewriting Notes

Notes should be rewritten as soon as possible after a lecture while the material is still fresh in your mind. In rewriting, the proper relationship in the notes can be clarified if necessary.

In addition to clarification of facts and relationship the rewriting helps you to remember the information since you must think about it then.

Drawings

Molds and yeasts are easily illustrated by making drawings of these microorganisms. In fact, these microorganisms may be presented in lectures as a combination of description in words and illustrations. A drawing of a filamentous fungus showing the production of conidia is called a habit sketch. It is essential that the student verify the appearance of the organisms by observations in the laboratory.

Life Cycles

The microorganisms (some molds and yeasts) which have a different morphological structure at different stages in their existence are easily illustrated using a combination of drawings and words and should be studied in this manner.

Daily Study

As in other sciences, food microbiology builds upon what one learns daily in terms of vocabulary, principles and examples. The material should be studied each day to integrate new knowledge and to repeat the information numerous times so that it will be retained. Only after a considerable time of study will the student have enough knowledge at hand for problem solving.

Spelling of Technical Words

The ability of the student to spell technical words is part of the course and part of the daily study routine. This skill is especially important in the taking of examinations, writing term papers, theses, etc.

Vocabulary

Each discipline has its own vocabulary. Food microbiology is no exception.

In thinking, speaking and writing in the science of microbiology, as in other disciplines, the learning of the proper nomenclature is an essential part of participating.

Reading Texts and Technical Papers

In studying a subject, there is an advantage in consulting several texts and technical papers on specific topics since various authors take different points of view. Recent technical papers will give the latest data and examples on specific subjects and are, therefore, ways to expand one's knowledge.

REVIEW

Answering Review Questions at the End of Chapters

The review questions are of the essay type which will be discussed below.

These questions follow the text and should give the student an adequate review of the text material.

Answer the question and then consult the text material to see how well you have done. Consult other texts to determine if any more could be added to your answer.

Making Your Own Review Questions

Make your own review questions as you study the text material, listen to the lectures, take notes and do additional reading. You can rephrase or reword the review questions or formulate entirely new questions. In doing this mental exercise, you should obtain a better understanding of the material.

Review of Previous Examinations

Examinations are just a way of sampling a person's knowledge on a subject. Exams are much more than a way to establish a grade; they are a study guide and teaching aid, too. If one can answer all conceivable questions put on the subject, then he knows the material.

EXAMPLES OF EXAMINATIONS

Essay Examinations

1. **What Should They Test?** Essay examinations should test how a person organizes his material, how he approaches the problem, how original and creative the person is, how well he understands principles and relationships and, lastly, how well he comprehends the entire subject matter.

2. **What Are the Key Words in the Question and/or Commands?** Key words give you direction in setting up your answer. For example, consider the following key words: compare, contrast, discuss, describe, define, effect, influence, explain, how, what, outline, list, why, give and name.

The key words which are used in commands are defined below:

>   *compare*—to show similarities or differences.
>   *contrast*—to compare so as to point out the differences.
>   *discuss*—to consider and to argue the pros and cons of an issue.
>   *describe*—to give a detailed account or to trace the outline of an issue.

define—to determine boundaries or limits; to give the distinguishing characteristics.
list—to enumerate.
outline—to give the essentials.
give—to list.
name—to list.

3.   **Outline the Answer First.** By outlining your answer before you start to write, you can give the proper order and relation to different items in your answer. Relationships are more easily seen by this method, especially the overview of the entire answer.

4.   **How Much Should Be Written?** The amount to be written depends upon the time and the knowledge of the person answering the question. The answer to a question during a 1 hr examination has to be shorter because of the restraints of time than the response to a question on a qualifying or comprehensive examination for a Ph.D. degree. In any case, the subject should be developed as completely as time will permit.

5.   **Generalizations and Specific Examples.** Usually essay examinations deal with principles. In answering questions dealing with principles, specific instances are necessary for illustration. The number of examples will depend upon the time available and the number known. Short answers are given in objective tests.

6.   **Advantages and Disadvantages.** The advantages of the essay examination are that the answerer is able to show the depth and scope of his knowledge, not just an isolated fact or two. Furthermore, such a test brings into play the ability to organize and communicate. Another major advantage is that the essay type lends itself to the testing of the ability to solve food microbiological problems.

The major disadvantage in an essay examination is that it limits the material upon which questions can be asked.

7.   **Value as Teaching and Learning Aids.** Since the essay examination samples knowledge in a given area, as a learning aid, it is also limited in scope. The essay examination gives more experience in the use of language and expression than does the objective type of test and it has, therefore, a broader value.

Objective Examinations

1.   **What Should They Test?** Objective type examinations test specific points rather than stressing broad concepts or principles.

2.   **Advantages and Disadvantages.** The answer is restricted in scope. Although the student uses little knowledge to answer a particular question, many questions can be asked per examination period. Organization of

material is not involved since the examination itself is already organized before the examination is given.

3.   As Teaching and Learning Aids. Objective tests can be used as teaching and learning aids if the teacher takes time to discuss the questions missed by the students.

4.   Types of Objective Questions.

True or False: The best type of true or false question is one for which the answerer must know multiple facts rather than only one fact. However, the information which makes a statement false should be a major point rather than something insignificant to be a fair question and to serve as a teaching and learning tool.

Fill in the Blank: This type requires a short answer and should be related to the important points of the course.

Matching: This type of question calls for an understanding of relationships which may be very direct or subtle. The more subtle the question, the more difficult it is to answer. Some words or phrases that cannot be matched should be placed in the question to reduce the chance of guessing.

Identification of Structures and Life Cycles: This is mainly a pictorial type of examination which may be by recognition (the drawings are provided) or by recall (the students draw the structure). Although microorganisms which have life cycles are few in number in the area of food microbiology, they are emphasized because of their classical importance.

Definitions: Vocabulary is a must in a science; hence, the ability to define terms is essential to an understanding of the subject. Usually, definitions of terms are short in length and are easily incorporated into objective type questions.

Short Answers: As stated above, definitions are a type of short answer. There are others in which only a word or two answer the question. This, too, allows rapid grading of the test.

Multiple Choice: With increased numbers of choices, the probability of guessing is reduced. With appropriate keys such examinations are easily and rapidly graded.

ESSAY AND OBJECTIVE PRACTICE EXAMINATION QUESTIONS COVERING SELECTED CHAPTERS

(Answers Given on Pages 266-270)

Students who were surveyed according to their likes or dislikes of the review questions at the end of the foregoing chapters found them helpful. Half of the students suggested that objective tests would be useful, too. This section was added to aid this group. For those students who do not have

TABLE 18.1.   EXAMPLES OF ANSWERS TO SELECTED ESSAY-TYPE QUESTIONS

Selected Subject	Question	Where Answer Is Found in Text (Pages)
Yeasts	*Discuss* the occurrence and importance of yeasts.	33–34
Molds	*How* can you preserve molds?	73 (short answer)
Factors influencing growth of spoilage microorganisms	*Why* do microorganisms need a source of N and C?	77–78
Microbiology of fish and seafood	*Compare* the glycogen in muscle of fish, beef, and poultry after slaughter. What influence does glycogen have on spoilage?	104–105, 120–121, 128–129
Microbiology of fermented foods	*Outline* steps in the production of beer.	195–196

an opportunity to take essay examinations as a practice exercise, such questions have also been included.

Selected review questions are listed in Table 18.1 which gives page numbers of the text material where acceptable assay-type responses are found. This method was used so that text material was not repeated.

Objective test questions were selected from tests previously given in the author's course in food microbiology.

Yeasts (Chapter 3)

True or False. If the sentence is completely true, mark the answer true (T). If all or part of the sentence is false, mark the answer false (F):

_____a.   Yeasts are found on plants, in soils, and in animals.

_____b.   Yeasts are pathogenic and saprophytic.

_____c.   Osmophilic yeasts are isolated easily on nutrient agar.

_____d.   Yeasts require C and N to grow.

_____e.   Asporogenous yeasts produce ascospores.

_____f.   Fermentation and assimilation mean the same thing.

_____g.   Most yeasts grow at 37°C (98.6°F).

_____h.   *Schizosaccharomyces* reproduces by budding whereas *Saccharomyces* reproduces by fission.

_____i.   Yeasts are classified according to morphological and physiological characteristics.

_____j.   Yeasts are a good source of protein.

Molds (Chapter 4)

Fill in the Blank. Add the word or phrase which will make a correct statement, using the information from the chapter.

1. _____ attacks potato and tomato plants. This fungus caused the blight in Ireland during 1845–1846.

2. _____ and _____ are two fungi known to produce amylases.

3. _____ is the fungus used in the production of tempeh.

4. _____, _____and _____ may be added to media to isolate fungi.

5. Fungi which produce ascospores are called _____; whereas fungi not producing sexual spores are called _____.

6. Nonseptate mycelium is called _____.

7. A structure that contains gametes is a _____.

8. _____ is sexual reproduction from the female gamete alone.

9. The bulbous head terminating the conidiophore of *Aspergillus* is a _____.

10. A _____ is a diploid cell resulting from the union of two haploid cells.

Factors Influencing Growth of Spoilage Microorganisms (Chapter 5)

Matching. Match the word or phrase in Column B with a word or words in Column A. Each word or phrase in Column B may be used only once.

Col. A	Col. B
____ 1. N source	a. coliforms, *Leuconostoc, Lactobacillus*
____ 2. C source	b. $-\log [H^+]$
____ 3. Osmotic pressure	c. amino acid
____ 4. pH	d. sugars
____ 5. Biological structure	e. heavy syrups
____ 6. O-R potential	f. sodium thioglycollate
____ 7. Water activity	g. egg shell
____ 8. Coenzyme	h. 1.0 to 0.0 scale
____ 9. Vitamin	i. Mn
____ 10. Metabiosis	j. biotin

Identification of Structures and Life Cycles (Chapter 6—Fruits and Vegetables)

1. Draw the life cycle of *Monilinia fructicola*. Label the parts.
2. Draw conidia of *Rhizopus, Alternaria* and *Aspergillus*.

Red Meats (Chapter 7)

Definitions. Give the meaning of the following terms in your own words or those given in the glossary.

1. Antemortem
2. Quick cure
3. Aflatoxin
4. Coagulase-positive staphylococci
5. Mycotoxin
6. Potable water
7. Comminuted products
8. Country-cured hams
9. Trichinae
10. Cresols

Poultry Meats (Chapter 8)

Short Answers.

1. Does poultry meat make a good substrate for the growth of spoilage flora? Why?
2. What bacterium associated with poultry meat can cause infection?
3. How can growth of bacteria on fresh poultry be controlled?
4. What anaerobic bacterium (which causes food poisoning) occurs with the greatest frequency?
5. Name three bacteria associated with poultry meat which may cause food poisoning and/or infection.

Fish and Seafood (Chapter 9)

Multiple Choice. Select the phrase that makes the sentence correct.

_____ 1. Fresh fish are (a) bright in appearance; (b) eyes are sunken; (c) flesh is soft and limp; (d) flesh leaves bones easily when split.

_____ 2. The dominant spoilage flora of fish are (a) *Pseudomonas;* (b) *Flavobacterium;* (c) *Micrococcus;* (d) all of these.

_____ 3. The method of fishing has (a) nothing to do with the numbers of bacteria on the fish; (b) very little to do with the numbers of bacteria on the fish; (c) most to do with the numbers of bacteria on the fish; (d) something to do with, but is only part of the influence of numbers of bacteria.

_____ 4. Fish and seafoods spoil rapidly because of (a) the pH; (b) the vitamins; (c) the proteins; (d) all of these.

_____ 5. Chemical indicators of quality of fish and seafoods are (a) indole; (b) ammonia; (c) pH; (d) all of these.

Eggs and Egg Products (Chapter 10)

True or False.

_____ 1. The pH of egg white and yolk is higher than that of most foods.

_____ 2. Lysozyme, an inhibitor of Gram-negative bacteria, is found in egg white.

_____ 3. Riboflavin chelates cations.

_____ 4. Holes in egg shells contain a protein-like substance.

_____ 5. The dominant spoilage flora of shell egg are *Pseudomonas*, *Micrococcus* and *Flavobacterium*.

_____ 6. Dirty eggs should be washed prior to being broken.

_____ 7. Dirty eggs should be washed prior to storage as shell eggs.

_____ 8. Formic, acetic, lactic and succinic acids are indicators of egg quality.

_____ 9. A temperature of 62.7°C (145°F) for 3.5 min is used to pasteurize liquid whole eggs.

_____10. A heat treatment of 54.4°C (130°F) for 15 min thermostabilizes the egg.

Dehydrated Fruits, Vegetables and Cereals (Chapter 11)

Fill in the Blank.

1. The water activity is equal to _____ for pure water.

2. If a food product with an a_w of 0.60 is placed in an environment with an ERH (equilibrium relative humidity) of 95% _____ will _____ .

3. _____ are the first microorganisms to grow on dehy-

drated foods when the moisture increases enough for growth to occur.

4. _____ and _____ are two genera which occur with a high frequency in spoiling dehydrated foods.

5. *Saccharomyces rouxii* growing on prunes is an example of _____ yeast.

6. _____ are indicative of _____ raw materials and/or _____ during dehydration.

7. Fungi may grow in cereal products producing _____ which cause health problems.

Canned Foods (Chapter 12)

Matching. Match the word or phrase in Column B with words in Column A. Use one word or phrase in Column B only once with a word in Column A.

Col. A	Col. B
_____ 1. Flat-sour	a. Butyric anaerobe
_____ 2. *B. coagulans*	b. PA3679
_____ 3. *D. nigrificans*	c. *D* value
_____ 4. Thermophilic an-aerobe	d. *F* and *z* values
	e. A mold that spoils canned food
_____ 5. *Clostridium botuli-num*	f. *B. stearothermophilus*
_____ 6. *Penicillium striatum*	g. *C. thermosaccharolyticum*
_____ 7. Thermal death time parameters	h. Food poisoning
	i. Tomato products
_____ 8. Rate of destruction parameters	j. Sulfide stinker
_____ 9. *Clostridium sporo-genes*	
_____ 10. *Clostridium butyri-cum*	

Chemically Preserved Foods and UV (Chapter 13)

Definitions. Define the following terms.
1. Jelly
2. Essential oil
3. Sorbic acid

4. Eugenol
5. Allyl isothiocyanate
6. Acrolein
7. Isobutylthiocyanate

Fermented Foods (Chapter 14)

Short Answers.
1. Give two functions of salt in kraut fermentation.
2. List the three species of bacteria in the final stage of the kraut fermentation.
3. What is malt?
4. What is mashing?
5. What is rye whiskey?
6. How many types of fermentation are there in manufacturing vinegar from apple juice?
7. Name two yeasts (other than *S. cerevisiae*) that are food yeasts.

Food Poisoning and Mycotoxins (Chapter 15)

Multiple Choice. Match the word or phrase in Column B with a word in Column A. Only one word or phrase in Column B is used twice.

Col. A	Col. B
_____ 1. *C. perfringens*	a. Causes most death in home-canned foods.
_____ 2. *C. botulinum*	b. Produces skin irritations.
_____ 3. *B. cereus*	c. Cows eat B_1 and excrete M_1.
	d. *Penicillium expansum.*
_____ 4. Aflatoxin	e. *C. botulinum.*
_____ 5. Alpha toxin	f. Carcinogenic to trout.
	g. Associated with poultry meats.
_____ 6. Type A to G	h. Facultative anaerobe that causes food poisoning.
_____ 7. Ochratoxin	i. A fungal toxin (general term).
_____ 8. Patulin	
_____ 9. T-2 toxin	
_____ 10. Mycotoxin	

ANSWERS TO OBJECTIVE TEST QUESTIONS GIVEN ON PAGES 259–265

Yeasts (Chapter 3)

T	a.	F	f.
T	b.	F	g.
F	c.	F	h.
T	d.	T	i.
F	e.	T	j.

Molds (Chapter 4)

1. *Phytophthora infestans*
2. *Aspergillus oryzae, Mucor rouxii, Rhizopus delemar* or *Rhizopus oligosporus*
3. *Rhizopus oligosporus*
4. Oxgall, crystal violet, and Rose bengal
5. Ascomycetes, Fungi Imperfecti
6. Coenocytic
7. Gametangium
8. Parthenogenesis
9. Vesicle
10. Zygote

Microorganisms (Chapter 5)

1.	c
2.	d
3.	e
4.	b
5.	g
6.	f
7.	h
8.	i
9.	j
10.	a

Identification of Structures and Life Cycles (Chapter 6)

Answer to assignment No. 1

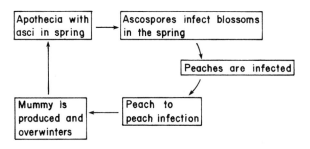

Answer to assignment No. 2

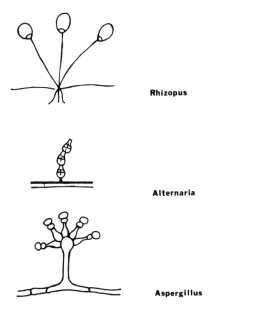

Red Meats (Chapter 7)

1. *Antemortem.* Before death.
2. *Quick cure.* The pickle contains less salt so the meat product must be refrigerated.
3. *Aflatoxin.* A toxin produced by several fungi. Some strains of *A. flavus* produce aflatoxin. There are several kinds of aflatoxin.

4. *Coagulase-positive staphylococci.* Staphylococci that produce the enzyme coagulase that coagulates blood serum.
5. *Mycotoxin.* A toxin produced by fungi.
6. *Potable water.* Water that is clean and fit to drink.
7. *Comminuted products.* Pulverized products. In this case, meat products.
8. *Country-cured hams.* Hams that are treated with high salt content in cure (with other chemicals) and are aged for 30 days or more. Do not require refrigeration.
9. *Trichinae.* Small worms whose larvae infest intestines and muscles of man, causing trichinosis. Can be caused by eating improperly cooked pork.
10. *Cresols.* Compounds which occur in smoke and are therefore found in smoked meats.

Poultry Meats (Chapter 8)

1. Yes, because of the pH and the nutrients which are available.
2. *Salmonella*
3. Sanitation, chlorination and proper refrigeration
4. *Clostridium perfringens*
5. *C. perfringens, Staphylococcus aureus, Salmonella* sp.

Fish and Seafood (Chapter 9)

a	1.
d	2.
d	3.
d	4.
d	5.

Eggs and Egg Products (Chapter 10)

T	1.		T	6.
F	2.		F	7.
T	3.		T	8.
T	4.		F	9.
F	5.		T	10.

Dehydrated Fruits, Vegetables and Cereals (Chapter 11)

1. 1.0.
2. *moisture* will *increase*
3. Molds
4. *Aspergillus* and *Penicillium*
5. an osmophilic
6. *High counts* / *poor* / *poor sanitation*
7. mycotoxins

Canned Foods (Chapter 12)

f	1.		e	6.
i	2.		d	7.
j	3.		c	8.
g	4.		b	9.
h	5.		a	10.

Chemically Preserved Foods and UV (Chapter 13)

1. *Jelly.* A mixture of fruit juice, sugar, and pectin adjusted to form a gel.
2. *Essential oil.* Volatile oils found in plants and/or spices.
3. *Sorbic acid.* A fungistat used in apple juice, bread and/or cheese to prevent mold growth.
4. *Eugenol.* An antimicrobial compound found in cloves.
5. *Allyl isothiocyanate.* An antimicrobial compound in mustard.
6. *Acrolein.* A compound having antimicrobial properties in garlic and onions.
7. *Isobutylthiocyanate.* A compound having antimicrobial properties in horseradish.

Fermented Foods (Chapter 14)

1. Salt causes a brine to form; salt is selective and allows lactic acid bacteria to dominate in the fermentation.
2. *Lactobacillus plantarum, L. brevis,* and *L. fermenti.*
3. Malt is an enzyme preparation made by germinating barley.
4. Mashing is an enzymatic process in which starch is hydrolyzed to fermentable sugars and some proteins are degraded to amino acids for food for the yeasts.

5. Rye whiskey is whiskey made from a mash consisting of 51% rye.
6. Two—alcoholic (anaerobic) and acetic (oxidative).
7. *Candida utilis* and *Candida tropicalis.*

Food Poisoning and Mycotoxins (Chapter 15)

g	1.		e	6.
a	2.		f	7.
h	3.		d	8.
c	4.		b	9.
g	5.		i	10.

POTPOURRI OF EXAMINATION QUESTIONS ON FOOD MICROBIOLOGY

(Answers Given on Pages 292–298)

Not all answers to the questions are discussed in this book. Students should consult various authors to increase their knowledge of each subject.

True or False

_____ **1.** Slow freezing does not injure bacteria as fast freezing does.

_____ **2.** Fresh fish have a backbone which is pearly gray in color; stale fish have a backbone which has a pink discoloration.

_____ **3.** Vitamins function as coenzymes in the metabolism of microorganisms.

_____ **4.** The yeast cell is a thallus.

_____ **5.** The decomposition of proteins to ammonia is called ammonification.

_____ **6.** Bacteria die while frozen because they starve to death, not because the freezing process injures the bacteria.

_____ **7.** Spray drying at 121.1°–204.4°C (250°–400°F) (hot air) sterilizes the food.

_____ 8. Fish decompose first along the tips of the ribs.

_____ 9. The odor of ammonia is indicative of decomposition in shrimp.

_____ 10. Proteolysis is a term used to describe the decomposition of ammonia to nitrogen and hydrogen.

_____ 11. Mycelium is a group of hyphae.

_____ 12. Members of Ascomycetes are coenocytic.

_____ 13. Protoplasmic streaming can be observed in some fungi.

_____ 14. *Rhodotorula* is a sporeforming yeast.

_____ 15. Osmotic pressure is equal to 12.04 Δ where Δ is the depression of the freezing point.

_____ 16. Although it works, the law forbids hydrogen peroxide sterilization of milk.

_____ 17. If a food contains thiamin, it can be preserved by sulfur dioxide.

_____ 18. Freeze-drying sterilizes foods.

_____ 19. Gas production always accompanies acid production.

_____ 20. A pink oyster is a spoiled oyster.

_____ 21. There are seven types of *Clostridium botulinum*.

_____ 22. The most common types of *C. botulinum* are A, B, and E in man.

_____ 23. Type E grows at higher temperatures than A or B.

_____ 24. Some filamentous fungi can utilize nitrate nitrogen.

_____ 25. The ammonium ion can be utilized by practically all yeasts.

_____ 26. Glucose is an amino acid which is utilized by more yeasts than any other amino acid.

_____ 27. Molds like a pH of 7 to 9 to grow and are, therefore, the main microbial contaminant of meats.

_____ 28. Sclerotia are holdfasts of *Rhizopus nigricans*.

_____ 29. Conidia are formed in a sporangiole.

_____ 30. Chlamydospore is another name for arthrospore.

_____ 31. Rhizomorphs are found in the genus *Rhizopus.*

_____ 32. One mole of any undissociated electrolyte in a liter of water exerts an osmotic pressure of 22.4 atmospheres.

_____ 33. Color defects by bacteria in milk are common.

_____ 34. Brick cheese is rarely contaminated by *Brevibacterium linens.*

_____ 35. Bacteria produce flavor in curing brines as used today.

_____ 36. Hydrogen gas produces eyes in Swiss cheese.

_____ 37. *Propionibacterium shermanii* produces hydrogen gas.

_____ 38. Aflatoxin causes more damage in young animals than in old ones.

_____ 39. Ergotism is a mycotoxicosis.

_____ 40. Aflatoxins can be destroyed easily by heat.

_____ 41. Infected individuals and fecal contamination cause most viral food poisoning outbreaks.

_____ 42. Type E *Clostridium botulinum* is the most heat-resistant type.

_____ 43. Type A *C. botulinum* is the common type found in fish products.

_____ 44. Eggs are a source of *Salmonella.*

_____ 45. *Salmonella* can survive the drying process and may be found in dried eggs.

_____ 46. Rats may harbor *Salmonella.*

_____ 47. Streptococci may cause food infection.

_____ 48. Both alpha and beta hemolytic streptococci have been involved in food infection.

_____ 49. Psychrophilic yeasts grow at 45°C (113°F).

_____ 50. Eh is a measure of the hydrogen ion concentration.

_____ 51. Anaerobic bacteria grow well on agar slants at 55°C (131°F).

_____ 52. Facultative anaerobes grow on agar slants.

_____ **53.** Foods may have Eh values that are positive and some negative.

_____ **54.** Osmotic pressure is a measure in pounds per square inch.

_____ **55.** An oogonium is a female reproductive body whereas an antheridium is a male reproductive structure.

_____ **56.** *Rhizopus nigricans* is an ascomycete which produces a zygospore.

_____ **57.** *Monilinia fructicola* produces perithecia.

_____ **58.** Part of the killing of bacteria due to freezing may be due to osmotic pressure effects.

_____ **59.** Vegetative cells are not affected by lyophilization.

_____ **60.** Both reproductive and physiological death may occur in microorganisms due to freezing.

_____ **61.** An a_w of 0.7 means the same percentage of water in any food.

_____ **62.** Two to five ppm chlorine are used in processing plants to reduce spoilage.

_____ **63.** The thermal death point of conidia of *Rhizopus nigricans* is 60°C (140°F).

_____ **64.** The more bacteria present in a system, the longer it takes to kill these bacteria with a lethal agent.

_____ **65.** Dehydrated fruits can contain more moisture than dehydrated vegetables without spoilage.

_____ **66.** A plasmolyzed cell is a cell in which the cytoplasm is shrunken.

_____ **67.** Cells usually plasmolyze in an isotonic solution.

_____ **68.** Eh means the oxidation-reduction potential.

_____ **69.** *Erwinia carotovora* is the most important bacterium causing spoilage in vegetables.

_____ **70.** Gray mold is *Rhizopus.*

_____ **71.** *Phytophthora infestans* caused the Irish migration to the United States.

_____ 72. The life cycle of a fungus may be divided into a vegetative phase and a reproductive phase.

_____ 73. The Eumycetes or false fungi are divided into four classes.

_____ 74. Hyphae are a group of mycelia.

_____ 75. Sclerotia and rhizomorphs are special spores.

_____ 76. *Aspergillus* means smooth-headed.

_____ 77. *Penicillium* means rough-headed.

_____ 78. A chlamydospore differs from an oidium in that the oidium is a thick-walled spore.

_____ 79. The genus *Alternaria* produces spores in clusters.

_____ 80. Yeasts reproduce by budding and fission.

_____ 81. Yeasts are classified either as Fungi Imperfecti or Ascomycetes.

_____ 82. Botulus in German means a sausage.

_____ 83. *Clostridium botulinum* causes a food infection.

_____ 84. *Staphylococcus* causes an intoxication.

_____ 85. Oxgall restricts the growth of molds.

_____ 86. *Bacillus coagulans* and *Bacillus stearothermophilus* are thermophiles and are aerogenic.

_____ 87. Autotrophic bacteria cause food spoilage.

_____ 88. Psychrophilic bacteria cause spoilage in canned foods.

_____ 89. No gas in 24 hr but gas in 48 hr in lactose broth indicates a doubtful presumptive test.

_____ 90. *Escherichia coli* colonies on EMB plates have a green metallic sheen.

_____ 91. There are fecal and nonfecal coliform bacteria.

_____ 92. A "swell" (spoiled canned food) should be flamed prior to opening.

_____ 93. The National Canners Association has standards for flat-sour spores in sugar.

_____ 94. *Desulfotomaculum nigrificans* is a mesophile which produces hydrogen sulfide.

_____ **95.** The Howard mold count method is used to judge the quality of tomato products.

_____ **96.** A high mold count indicates either unsanitary conditions of manufacture or unfit raw materials or a combination of both.

_____ **97.** Microaerophiles require a limited oxygen tension.

_____ **98.** A positive Eh indicates anaerobic conditions.

_____ **99.** There are gradations of Eh values which influence the growth of anaerobes.

_____**100.** Xerophilic means dryness-loving.

_____**101.** Acetic acid preserves better than citric acid.

_____**102.** Chlamydospores are thin-walled spores.

_____**103.** Budding may be dipolar or multipolar.

_____**104.** *Pichia* and *Hansenula* are contaminants in the brewing industry.

_____**105.** *Flavobacterium* and *Micrococcus* are Gram-negative.

_____**106.** Moisture on the surface of meat adds to the potential of spoilage.

_____**107.** Aerobic decomposition is called putrefaction.

_____**108.** *Erwinia carotovora* produces pectolytic enzymes.

_____**109.** Osmophilic yeasts spoil honey.

_____**110.** Bacteria are more easily killed by heat in an acid food.

_____**111.** *Schizosaccharomyces* are yeasts that reproduce by budding.

_____**112.** Molds can grow inside shell eggs.

_____**113.** Toxin of *C. botulinum* is heat stable.

_____**114.** *Bacillus stearothermophilus* is a mesophile.

_____**115.** *Clostridium sporogenes* is highly saccharolytic.

_____**116.** *C. botulinum* is a nonsporeforming mesophile.

_____**117.** Mother of vinegar is growth of bacteria in vinegar.

_____**118.** Yeasts produce vitamins.

_____119. Both karogamy and plasmogamy occur in the life cycle of *Schizosaccharomyces octosporus*.

_____120. *Cryptococcus* is a sporeforming yeast.

_____121. False yeasts form spores.

_____122. In the fermentation of sauerkraut and cucumbers, the dominant flora are the same.

_____123. Agar starts to gel at 7.2°C (45°F).

_____124. *Pseudomonas, E. coli* and *Bacillus coagulans* are Gram-negative rods.

_____125. In counting plates to enumerate the number of bacteria, one should count 30–300 colonies because this gives the best results statistically.

_____126. Pipettes can be sterilized by heating them at 130°C (266°F) (dry heat) for 5 to 6 hr.

_____127. In considering the shape and the manner of cell division, there are four kinds of cocci.

_____128. Rough and smooth are terms used to describe the appearance of bacterial colonies.

_____129. Amphitrichous means that flagella are present on all sides of the bacterial cell.

_____130. The capsule in the genus *Bacillus* is a polypeptide.

_____131. Lysozyme breaks a hexosamine-containing mucocomplex or mucopolysaccharide in the cell wall.

_____132. Bacterial spores may be dormant.

_____133. Conidia of *Rhizopus* are produced in a sporangium whereas the conidia of *Aspergillus* are produced on sterigmata.

_____134. Oidia are thick-walled spores occurring in the mycelium whereas chlamydospores are thin-walled spores which break off from the tips of hyphae.

_____135. *Penicillium* has a foot cell which makes it easily distinguished from the rhizoids of *Rhizopus*.

_____136. Two characteristics of the Fungi Imperfecti are coenocytic mycelium and the nonseptate mycelium.

_____137. Sporangioles are immature sporangia.

_____138. Antheridia are female reproductive cells.

_____139. An apothecium is the same as a cleistothecium.

_____140. Osmophilic yeasts can grow at lower available water content than can halophilic bacteria.

_____141. Aerobes can grow and reduce the Eh to a level where anaerobes can grow.

_____142. *Bacillus coagulans* and *Clostridium butyricum* can grow at a pH below 4.5.

_____143. Molds can grow in general at a more acid pH than bacteria do in general.

_____144. "Sweating" of dry fruits to equalize the moisture content may permit some microbial growth.

_____145. Blanching and treatment with sulfur dioxide of vegetables can reduce the number of microorganisms on the food.

_____146. Dehydrated vegetables usually have less moisture than dehydrated fruits.

_____147. *Clostridium botulinum* type A has the most heat-resistant spores.

_____148. *Pseudomonas* predominates on shrimp during prolonged storage time in crushed ice.

_____149. *Clostridium perfringens* causes more food poisoning than *C. botulinum*.

_____150. Acetylmethylcarbinol may be used as an indicator of quality in an apple juice processing plant.

_____151. The presence of diacetyl in apple juice is an indication of poor sanitation in an apple juice processing plant.

_____152. Sorbic acid inhibits bacterial growth by influencing the cell membrane.

_____153. *Alternaria solani* is a yeast that spoils alcohol.

_____154. A film yeast is the same thing as a top yeast.

_____155. *Staphylococcus* is a test fungus used in canning.

_____156. The lag period for psychrophiles is shorter at low temperatures than for mesophiles.

_____157. Biological reactions do not follow the rule (Q_{10} = 2 to 4) except approximately in the middle of their active temperature range.

_____158. The generation time of a psychrophile and mesophile may be very close at 30°C (86°F) but far apart at 5°C (41°C).

_____159. In decomposing chicken meat, odor develops before slime appears.

_____160. *Pseudomonas* and *Alcaligenes* dominate in the spoilage flora at 4°C (39.2°F) on chicken meat.

_____161. *Shigella histolytica* causes amoebic dysentery.

_____162. The physiological condition of beef cattle prior to slaughter has no influence on the storage life of meat.

_____163. It is recommended to wet the surface of animal carcasses prior to and during chilling to prevent dehydration of the meat.

_____164. Clostridia have active catalase enzymes which break down H_2O_2 in the cell.

_____165. A summer sausage is a fermented product.

_____166. Carbon dioxide suppresses *Pseudomonas* and *Proteus* to prevent their growth on meat.

_____167. Flat-sour spores may accumulate in a blancher during periods when the plant is not operating.

_____168. A thermal death time curve is a plot of the log of time vs temperature in degrees F.

_____169. Acetic acid has a lower pH at which microorganisms are inhibited than does lactic acid.

_____170. Eugenol is the active antimicrobial ingredient in garlic.

_____171. Nisin is destroyed by the gastric juices.

_____172. Physiological death precedes reproductive death.

_____173. Osmotic pressure may kill a cell during slow freezing.

_____174. Temperatures higher than −10°C (14°F) probably are more lethal to bacteria because the processes of protein denaturation are less at lower temperatures.

_____175. Enterococci are more resistant to freezing and storage than coliforms.

_____176. Available moisture is equal to the vapor pressure of the solution or food divided by the vapor pressure of the solvent (water).

_____177. *Salmonella* may be spread by rats.

_____178. The toxin of *Staphylococcus* is more resistant to heat than the toxin of *C. botulinum*.

_____179. *Staphylococcus* does not grow above 48.8°C (120°F) nor below 4.4°C (40°F).

_____180. *Salmonella* produces an infection whereas *Staphylococcus* produces an intoxication.

_____181. Liquid whole eggs are pasteurized at 60°C (140°F).

_____182. Some yeasts produce pseudomycelium.

_____183. Ammonia may be given off during the breakdown of proteins by bacteria.

_____184. *Aspergillus* has a vesicle whereas *Penicillium* does not.

_____185. Fungi classified as Fungi Imperfecti produce sexual spores.

_____186. Foods with a pH of 4.8 should be canned using a temperature greater than 100°C (212°F).

_____187. Nitrate can be reduced by bacteria to form nitrite in meat products.

_____188. *Monilinia fructicola* causes brown rot of stone fruits.

_____189. The most germicidal wavelengths of ultraviolet light are 2650–2660 Å.

_____190. Ozone may be used to suppress microbial growth.

_____191. The production of vinegar is an example of metabiosis.

_____192. Chickens should not be fed prior to slaughter to maintain better sanitation during slaughter.

_____193. *Lactobacillus* can cause greening in sausage.

_____194. *Bacillus stearothermophilus* causes flat-sour spoilage in tomato juice.

_____195. Aspergilli and penicillia are found on aging country-cured hams.

_____196. Sodium nitrite is bacteriostatic.

_____197. The accuracy of a direct microscopic count is dependent on the differential stainability of the cells and the number of them counted.

_____198. Alpha toxin of *Staphylococcus aureus* produces clear zone hemolysis on blood agar.

_____199. Streptococci require more growth factors than do the pseudomonads.

_____200. Homofermentative lactobacilli include the species *bulgaricus, casei* and *plantarum*.

Matching

Preservatives

Match the words or phrases in Column A with the words or phrases in Column B. Use all phrases, symbols in Column B only once.

Col. A	Col. B
201. _____ Cinnamon	a. nicotinic acid interferes with its inhibition
202. _____ Mustard	
203. _____ Garlic & onions	b. ties up acetaldehyde
204. _____ Horseradish	c. interferes with the synthesis of pantothenic acid
205. _____ Cloves	
206. _____ Sulfur dioxide	d. interferes with the dehydrogenase system
207. _____ Propionic acid	
208. _____ Sorbic acid	e. 2660 Å
209. _____ Benzoic acid	f. cinnamic aldehyde
210. _____ Ultraviolet light	g. allyl isothiocyanate
	h. eugenol
	i. acrolein
	j. isobutylthiocyanate

Fermented Foods

Match the words or phrases in Column A with the words or phrases in Column B. Some are used more than once.

Col. A	Col. B
211. _____ Kraut	a. alcohol
212. _____ *Saccharomyces cerevisiae*	b. homo- and heterofermentative bacteria

	Col. A		Col. B
213.	_____ *Acetobacter aceti*	c.	food yeasts
214.	_____ Alpha amylase	d.	ferments lactose
215.	_____ Beta amylase	e.	malt
216.	_____ Ferments cucum- bers	f. g.	aging salt stock
217.	_____ Ester production	h.	aerobic
218.	_____ Acetic acid produc- tion	i. j.	anaerobic fruit juices
219.	_____ Alcohol production	k.	corn
220.	_____ Brandy	l.	sloe berry
221.	_____ Bourbon	m.	beer
222.	_____ Sloe gin	n.	*Enterobacter aerogenes*
223.	_____ Mashing	o.	pink kraut
224.	_____ Floaters		
225.	_____ *Rhodotorula*		

Food Poisoning

Match the word or phrase in Column A with words or phrases in Column B.

	Col. A		Col. B
226.	_____ *C. perfringens*	a.	patulin
227.	_____ *C. botulinum* type A	b.	aflatoxin
228.	_____ *B. cereus*	c.	ochratoxin
229.	_____ *Aspergillus flavus*	d.	facultative anaerobe
230.	_____ *Penicillium expan-* *sum*	e. f.	$F = 2.78$ min alpha toxin
231.	_____ *Aspergillus ochra-* *ceus*	g. h.	marine foods M aflatoxin
232.	_____ *Aspergillus versi-* *color*	i. j.	sterigmatocystin seven types
233.	_____ *C. botulinum* type E	k.	most toxic toxin known
234.	_____ *C. botulinum*	l.	trypsin
235.	_____ B aflatoxin	m.	*C. perfringens*
236.	_____ *C. botulinum*	n.	*C. botulinum*
237.	_____ Activates botulinal toxin	o.	mycotoxins
238.	_____ Meats and gravies		
239.	_____ Home-canned foods		
240.	_____ Cereals		

Genus and Class

Match the word in Column A with a word in Column B. Each word in Column B may be used more than once.

Col. A		Col. B	
241.	_____ *Rhizopus*	a.	Deuteromycetes
242.	_____ *Phytophthora*	b.	Ascomycetes
243.	_____ *Alternaria*	c.	Oomycetes
244.	_____ *Hansenula*	d.	Zygomycetes
245.	_____ *Kloeckera*		
246.	_____ *Schizosaccharo-*		
	myces		
247.	_____ *Neurospora*		

pH and Food Groups

Match the spoilage microorganism in Column B with food groups (according to pH) in Column A. There are multiple answers to each word or phrase in Column A.

Col. A		Col. B	
248.	_____ Low-acid, group I	a.	*C. pasteurianum*
249.	_____ Medium-acid,	b.	*C. butyricum*
	group II	c.	*Lactobacillus brevis*
250.	_____ Acid, group III	d.	yeasts
251.	_____ High-acid, group	e.	molds
	IV	f.	*B. stearothermophilus*
		g.	*C. thermosaccharolyticum*
		h.	*D. nigrificans*
		i.	*C. sporogenes*
		j.	*B. licheniformis*
		k.	*B. cereus*

General Matching Set No. 1

Match the words or phrases in Column A with a word or phrase of Column B. Single answer per word or phrase in Column A.

Col. A		Col. B	
252.	_____ Flat-sour spoilage	a.	*Thamnidium*
253.	_____ Whiskers on aging	b.	*Pseudomonas*
	beef	c.	*F* and *D* values

	Col. A		Col. B
254.	_____ Dominant spoilage bacterium after spoilage of seafood at refrigeration temperatures	d.	2–5 ppm chlorine
		e.	ultraviolet light
		f.	canned food
		g.	film yeast
		h.	_C. sporogenes_
255.	_____ Thermal resistance		
256.	_____ Continuous sanitizing		
257.	_____ Controlling mold growth during aging of meat		
258.	_____ Spoilage of fermented cucumbers		
259.	_____ Canned food blowing up		

General Matching Set No. 2

Match the words or phrases in Column A with words or phrases of Column B. Single answer per word or phrase in Column A.

	Col. A		Col. B
260.	_____ Basidiospore	a.	basidium
261.	_____ _Cryptococcus_	b.	apothecium
262.	_____ Cuplike ascus-containing body	c.	cleistothecium
		d.	columella
263.	_____ _Rhizopus_	e.	sterigmata
264.	_____ _Penicillium_	f.	little brush
265.	_____ Available water	g.	DPT
266.	_____ Thiamin	h.	DPN and TPN
267.	_____ Shows uptake of water	i.	a_w
		j.	halophilic
268.	_____ Dry-loving	k.	xerophilic
269.	_____ Kraut is an example	l.	Eh
		m.	anaerobiosis
270.	_____ Salt-loving	n.	antibiosis
271.	_____ A means of reproducing in _Saccharomyces_	o.	metabiosis
		p.	sorption isotherms for vegetables
272.	_____ _Candida_	q.	budding

Col. A   Col. B

273. _____ A fruiting body   r. fission
containing asci and   s. reduction division
no special opening   t. karogamy
274. _____ Aspergillus   u. plasmogamy
275. _____ Nicotinic acid   v. legitimate diploid
276. _____ Kloeckera   w. apiculate yeast
277. _____ Life against life   x. film yeast
278. _____ Fusion of nuclei   y. true yeast
279. _____ Pairing of nuclei   z. false yeast
280. _____ Has four asco-
spores
281. _____ Schizosaccharomyces
282. _____ Meiosis
283. _____ Lack of oxygen
284. _____ Oxidation-reduction
285. _____ Saccharomyces

Fungi Matching

Match the word or phrase in Column A with words or phrases in Column
B. Single answer per word or phrase in Column A.

Col. A   Col. B

286. _____ Thallus   a. the bulbous head terminating
287. _____ Sporangium   the conidiophore of Aspergil-
288. _____ Zygote   lus.
289. _____ Sclerotium   b. a relatively simple plant body
290. _____ Gamete   devoid of stems, roots and
291. _____ Chlamydospore   leaves; in fungi the somatic
292. _____ Oidium   phase.
293. _____ Septate   c. a spore formed asexually at the
294. _____ Conidium   tip or side of a hypha.
295. _____ Vesicle   d. with cross walls.
                                  e. a diploid cell resulting from the
                                     union of two haploid cells.
                                  f. a sac-like structure, the entire
                                     protoplasmic contents of which
                                     become converted into an in-
                                     definite number of spores.

Col. B

g. a hard resting body resistant to unfavorable conditions, which may remain dormant for long periods of time and germinate upon return of favorable conditions.

h. a thin-walled, free, hyphal cell derived from fragmentations of a somatic hypha into its component cells.

i. a hyphal cell enveloped by a thick cell wall, which eventually becomes separate from the parent hypha and behaves as a resting spore.

j. a differentiated sex cell or a sex nucleus which fuses with another in sexual reproduction.

Media and Food Poisoning Matching

Match the word or phrases in Column A with words or phrases of Column B. Some words or phrases can have multiple answers.

Col. A		Col. B	
296.	_____ Staphylococcus 110 agar	a.	*Salmonella*
297.	_____ Mannitol salt agar	b.	7.5% NaCl
298.	_____ TEPY	c.	egg yolk
299.	_____ Endo agar	d.	polymyxin B
300.	_____ Brilliant green phenol red agar	e.	Dolman kitten test
301.	_____ Test for *Staphyloccus* toxin	f.	staphylococcal food poisoning
		g.	botulism
302.	_____ Most deaths	h.	*Staphylococcus*
303.	_____ Least deaths	i.	*Escherichia coli*
304.	_____ Spoils fish	j.	*Pseudomonas*

General Matching Set No. 3

Match the word or phrase in Column A with words or phrases in Column

B. Some words or phrases in Column A can have multiple answers.

	Col. A		Col. B
305.	_____ Whiskey	a.	*C. thermosaccharolyticum*
306.	_____ Yeast production	b.	preservation (usually)
307.	_____ Vinegar production	c.	microaerophilic
308.	_____ Kraut	d.	molds
309.	_____ Lactobacilli	e.	an anaerobic process
310.	_____ a_w of 0.7	f.	film yeast
311.	_____ Causes spoilage of alcohol in vinegar	g.	*Acetobacter*
		h.	an aerobic process
312.	_____ Thermophiles	i.	*D. nigrificans*
313.	_____ Causes spoilage of fermented cucumbers	j.	salt and lactic acid
		k.	*B. stearothermophilus*
		l.	*C. botulinum*
314.	_____ Causes spoilage of vinegar	m.	*C. sporogenes*
		n.	*C. butyricum*
315.	_____ Causes primary spoilage of dehydrated foods		
316.	_____ Preserves fermented cucumbers		
317.	_____ Mesophiles		

Multiple Choice

_____318. A spore may be (a) conidium; (b) endospore; (c) arthrospore; (d) chlamydospore; (e) all of these.

_____319. Endospores are formed by (a) *Bacillus;* (b) *Salmonella;* (c) *Enterobacter;* (d) all of these.

_____320. A bacterial endospore has (a) a multilayered coat; (b) a cortex; (c) a spore core; (d) all of these.

_____321. Germination may be accelerated by (a) alanine; (b) oleic acid; (c) linoleic acid; (d) none of these.

_____322. Spores which go into dormancy (at the normal temperature of growth) are said to be in (a) heat-shock; (b) heat-induced dormancy; (c) neither of these.

_____323. Bacteria causing spoilage of fish are (a) *Botrytis* and *Mucor;* (b) *Pseudomonas* and *Flavobacterium;* (c) *Acetobacter* and *Lactobacillus.*

_____324. The recommended moisture to prevent growth of microorganisms in dehydrated peaches is (a) 25–30%; (b) 4%; (c) 15–20%.

_____325. The usual temperature for the dehydration of fruits is (a) 60°–73.9°C (140°–165°F); (b) 85°–93.3°C (185°–200°F); (c) 57.2°–93.3°C (135°–200°F).

_____326. Sulfur dioxide, sodium sulfite, or persulfite (a) reduces S-S links in enzymes; (b) influences the capsule of bacterial cells; (c) neither of these.

_____327. The following microorganisms are commonly found on dehydrated fruit: (a) molds; (b) yeasts; (c) molds and yeasts; (d) molds, yeasts and lactobacilli.

_____328. High total counts in dehydrated foods indicate (a) poor raw materials; (b) improper processing techniques; (c) poor sanitation; (d) all of these.

_____329. Freezing may (a) change the pH of the food (0.3 to 2.0 pH units); (b) disturb colloids; (c) increase osmotic pressure by changing ionic strength; (d) produce all of these.

_____330. As a rule (a) yeasts and molds are less susceptible to freezing than are most bacteria; (b) bacteria and molds are less susceptible than yeasts; (c) bacteria and yeasts are less susceptible than molds.

_____331. Nonsporeforming bacteria in the (a) logarithmic growth phase; (b) log and intermediary phases are more susceptible to freezing.

_____332. (a) Sugar; (b) fat; (c) alginate; (d) all of these may protect microorganisms during freezing.

_____333. Metabolic injury of frozen cells may be determined by (a) a minimal medium; (b) complete and minimal media; (c) neither of these.

_____334. Death in bacteria is thought to be due to (a) ice crystals; (b) protein denaturation; (c) neither of these.

_____335. Several studies have indicated that more microorganisms are destroyed at (a) −4°C (24.8°F); (b) −15° to −24°C (5° to 11.2°F); (c) below −24°C (11.2°F) than at other temperatures.

_____336. E. coli is (a) Gram-negative; (b) Gram-positive; (c) Gram-variable.

————337. *Schizosaccharomyces pombe* reproduces asexually by (a) budding; (b) fission and arthrospores; (c) budding and ascospores; (d) fission.

————338. Plasmogamy occurs in *S. octosporus* (a) between ascospores; (b) between vegetative cells; (c) does not occur.

————339. *Candida* forms (a) mycelium; (b) mycelium and budding cells; (c) mycelium, budding cells, and ascospores.

————340. *Endomycopsis* forms (a) mycelium; (b) mycelium and budding cells (c) mycelium, budding cells and ascospores.

————341. A fungus (a) usually can grow at a lower a_w than bacteria; (b) grows at a higher a_w than bacteria; (c) grows at an intermediate a_w.

————342. A yeast cell which has been growing at a low a_w (for example 0.7) if placed in pure water (a) would plasmolyze; (b) would swell; (c) would not be influenced at all.

————343. Oxidation-reduction (O-R) potential of a food (a) is a factor influencing the growth of microorganisms; (b) is not significant in food microbiology; (c) is important only in medical microbiology.

————344. The O-R potential is measured (a) by instruments; (b) by dyes; (c) by dyes or instruments.

————345. Molds grow in an O-R potential of (a) positive values; (b) negative values; (c) only at zero values.

————346. A negative O-R potential may be produced by (a) aerobic bacteria growing first in an enclosed container; (b) using carbon dioxide gas; (c) both of these.

————347. Psychrophilic aerobic sporeforming bacteria (a) exist in tropical countries; (b) do not exist; (c) have been found in dairy products.

————348. Sporeforming aerobic rods (a) occur as psychrophiles; (b) occur as mesophiles; (c) occur as psychrophiles, mesophiles and thermophiles.

————349. Thermophiles are of primary concern to (a) the canning industry; (b) the frozen food industry; (c) the fermentation industry.

————350. The spores of *C. botulinum* type E are (a) more heat-resistant than yeast cells; (b) less heat-resistant than type B; (c) are the most heat-resistant of spores of cell types of *C. botulinum*.

_____**351.** The order of heat resistance is that (a) bacterial cells are more heat-resistant than yeast cells; (b) mold cells are more heat-resistant than yeast cells; (c) bacterial spores are more heat-resistant than bacterial vegetative cells, yeast cells and molds.

_____**352.** Fermented salt stock (cucumbers) are preserved by (a) acetic acid; (b) lactic acid; (c) lactic acid and 15% salt solution.

_____**353.** Yeasts reproduce by (a) budding; (b) fission; (c) budding and fission.

_____**354.** The yeast which reproduces by fission is (a) *Saccharomyces;* (b) *Schizosaccharomyces;* (c) *Rhodotorula.*

_____**355.** The following has a foot cell (a) *Penicillium;* (b) *Aspergillus;* (c) *Alternaria.*

_____**356.** *Thamnidium* has (a) sporangia; (b) sporangia and sporangioles; (c) neither of these.

_____**357.** You can tell *Mucor* from *Rhizopus* by the fact that (a) *Mucor* has stolons and *Rhizopus* does not; (b) *Rhizopus* has rhizoids and *Mucor* does not; (c) *Mucor* has club-shaped conidia and *Rhizopus* does not.

_____**358.** A zygospore is produced by (a) *Rhizopus;* (b) *Alternaria;* (c) *Botrytis.*

_____**359.** *Aspergillus* can be distinquished from *Penicillium* by the fact that (a) *Aspergillus* has a vesicle and *Penicillium* does not; (b) *Penicillium* has a vesicle and *Aspergillus* does not; (c) the spores of *Aspergillus* are produced in a sporangium and *Penicillium* spores are not.

_____**360.** The fungus *Botrytis* is characterized by the fact that spores are (a) produced in chains; (b) produced in grape-like bunches; (c) produced in a sporangium.

Exercises with Scientific Facts

Listing of Facts

361. List five chemical preservatives and give the probable mode of action of each.

362. List the dominant spoilage flora for beef and poultry.

363. Give the pH classification of foods and the spoilage microorganisms with different pH groups in canned foods.

364. Give one antibiotic which has been suggested to be used in canned foods.

Graphing of Facts

Draw graphs to illustrate the following facts. Label the ordinate and the abscissa; use appropriate labeling to make the graph clear and understandable to anyone.

365. The effect of pH on thermal resistance of bacterial spores and the effect of numbers of spores (in the same graph).

366. The comparative resistance of molds and yeasts to heat.

367. A thermal death time curve.

368. Compare the resistance of enzymes to spores of *C. botulinum* to heat and to ionizing radiation.

369. How handling methods influence the spoilage of canned foods.

370. The normal fermentation of kraut.

371. The influence of pH on the preserving power of benzoic acid and sorbic acid.

372. The growth of bacteria with time.

373. A graph illustrating the direct hit theory of killing yeasts with ionizing radiation.

Association of Facts

374. Give the significance of each of the following bacteria in foods.
  a. *Bacillus stearothermophilus*
  b. *Salmonella*
  c. *Staphylococcus*

375. Why does pH have an influence upon microbial growth?

376. In what foods are psychrophiles important as spoilage flora?

377. Give two foods in which high bacterial plate counts do not indicate spoilage or low quality.

378. Give the name of bacteria associated with:
  a. greening of sausage
  b. flat-sour spoilage in tomato juice

379. Give the active antimicrobial agent in the following:
  a. cloves
  b. garlic
  c. horseradish

380. How is sanitation related to the shelf-life of a food product?

Putting Facts to Work

381. Draw a flow sheet for the killing of beef. List the places where possible contamination may occur.

382. You have developed a new synthetic food product and you wish to preserve it. The gross chemical analysis is as follows: pH, 6.0; protein, 5%; carbohydrate, 18%; fat, 1%; ash, 2%; moisture, 74%.

a.   Tell how you would can it.
b.   Dehydrate it.
c.   Can any chemical preservatives be used? Why?

Fill in the Blanks

383.   The temperature that is used in incubation of samples in the Eijkman test is _____ °C.

384.   List five diseases that may be spread by cups, spoons, glasses, dishes, and discharges of nose and throat.

385.   A person who has disease-causing bacteria in his body and sheds them in his feces is called a _____.

386.   List five diseases that are transmitted by the vehicle of food.

387.   Give the scientific names for the bacteria which cause
          a.   typhoid fever
          b.   paratyphoid fever
          c.   amoebic dysentery
          d.   bacillary dysentery.

388.   The recommended time and temperature for sanitizing dishes by hand are: _____ °F for _____ min or _____ °F for _____ min.

389.   _____ parts per million of chlorine should be used to sanitize dishes after washing.

390.   If machine dishwashing is done, the recommended temperature for washing is _____ °F and for rinsing is _____ °F.

391.   In evaluating the degree cleanliness (bacterial), the plate count per utensil should not exceed _____ if the utensil is to be considered satisfactory.

392.   *Clostridium botulinum* produces an _____ toxin whereas *Salmonella* produces an _____.

Definitions

Define the following terms:
393.   ascus
394.   meiosis
395.   mitosis
396.   oogonium

397. diploid
398. heterothallic
399. sclerotium
400. sporangium

ANSWERS TO TEST QUESTIONS GIVEN ON PAGES 270-292

1.	f	36.	f	71.	t	106.	t
2.	t	37.	t	72.	t	107.	f
3.	t	38.	t	73.	f	108.	t
4.	t	39.	t	74.	f	109.	t
5.	f	40.	f	75.	f	110.	t
6.	f	41.	t	76.	f	111.	f
7.	f	42.	f	77.	f	112.	t
8.	t	43.	f	78.	f	113.	f
9.	t	44.	t	79.	f	114.	f
10.	f	45.	t	80.	t	115.	f
11.	t	46.	t	81.	t	116.	f
12.	f	47.	t	82.	f	117.	t
13.	t	48.	t	83.	f	118.	t
14.	f	49.	f	84.	t	119.	t
15.	t	50.	f	85.	t	120.	f
16.	t	51.	f	86.	f	121.	f
17.	f	52.	t	87.	f	122.	t
18.	f	53.	t	88.	f	123.	t
19.	f	54.	f	89.	t	124.	f
20.	t	55.	t	90.	t	125.	t
21.	t	56.	f	91.	t	126.	t
22.	t	57.	f	92.	f	127.	t
23.	f	58.	t	93.	t	128.	t
24.	t	59.	f	94.	f	129.	f
25.	t	60.	t	95.	t	130.	t
26.	f	61.	f	96.	t	131.	t
27.	f	62.	t	97.	t	132.	t
28.	f	63.	t	98.	f	133.	t
29.	f	64.	t	99.	t	134.	f
30.	f	65.	t	100.	t	135.	f
31.	f	66.	f	101.	t	136.	f
32.	t	67.	f	102.	f	137.	f
33.	f	68.	t	103.	t	138.	f
34.	f	69.	t	104.	t	139.	f
35.	t	70.	f	105.	f	140.	t

141. t	184. t	227. e	270. j
142. t	185. f	228. d	271. q
143. t	186. t	229. b	272. x
144. t	187. t	230. a	273. c
145. t	188. t	231. c	274. e
146. t	189. t	232. i	275. h
147. t	190. t	233. g	276. w
148. t	191. t	234. j or k	277. n
149. t	192. t	235. h	278. t
150. t	193. t	236. k or j	279. u
151. t	194. f	237. l	280. v
152. f	195. t	238. m	281. r
153. f	196. t	239. n	282. s
154. f	197. t	240. o	283. m
155. f	198. f	241. d	284. l
156. t	199. t	242. c	285. y
157. t	200. t	243. a	286. b
158. t	201. f	244. b	287. f
159. t	202. g	245. a	288. e
160. t	203. j	246. b	289. g
161. f	204. i	247. b	290. j
162. f	205. h	248. f,g,h,i,j,k	291. i
163. f	206. b	249. g,f,i,j	292. h
164. f	207. c	250. a,b	293. d
165. t	208. d	251. c,d,e	294. c
166. t	209. a	252. f	295. a
167. t	210. e	253. a	296. h,b
168. t	211. b	254. b	297. b,h
169. f	212. a,c,m	255. c	298. c,d,h
170. f	213. a,h,j	256. d	299. j,a,i
171. t	214. e,m	257. e	300. a
172. f	215. e,m	258. g	301. e
173. t	216. b	259. h	302. g
174. t	217. f	260. a	303. f
175. t	218. a,h	261. z	304. j
176. t	219. i	262. b	305. e
177. t	220. j	263. d	306. h
178. t	221. k	264. f	307. g,h
179. t	222. l	265. i	308. f,j,c
180. t	223. m	266. g	309. c,j
181. t	224. n	267. p	310. b
182. t	225. o	268. k	311. f
183. t	226. f	269. o	312. a,k,i

313.	f	325.	a	337.	d	349.	a
314.	g	326.	a	338.	b	350.	b
315.	d	327.	d	339.	b	351.	c
316.	j	328.	d	340.	c	352.	c
317.	g,l,m,n	329.	d	341.	a	353.	c
318.	e	330.	c	342.	b	354.	b
319.	a,b	331.	b	343.	a	355.	b
320.	d	332.	d	344.	c	356.	b
321.	a	333.	b	345.	a	357.	b
322.	b	334.	b	346.	c	358.	a
323.	b	335.	a	347.	c	359.	a
324.	c	336.	a	348.	c	360.	b

361. (1) Benzoate, destruction of membranes; (2) sulfites, reduction of S-S linkage in enzyme protein; (3) salt, precipitation of enzyme protein; (4) sorbic acid, interferes with dehydrogenase system; (5) propionic acid, competes with certain amino acids for space on enzyme groups.

362. Dominant flora of beef: *Pseudomonas, Flavobacterium, Micrococcus, Cladosporium* and *Thamnidium;* poultry: *Pseudomonas, Flavobacterium* and *Micrococcus.*

363. The pH classification and spoilage flora are as follows:

Group I, Low-acid: *B. stearothermophilus, C. thermosaccharolyticum, D. nigrificans, C. sporogenes, B. licheniformis,* and *B. cereus.*

Group II, Medium-acid: *C. thermosaccharolyticum, B. stearothermophilus, C. sporogenes, D. nigrificans.*

Group III, Acid: *C. pasteurianum* and *C. butyricum*

Group IV, Very Acid: *Lactobacillus,* yeasts and molds.

364. Nisin.

365.

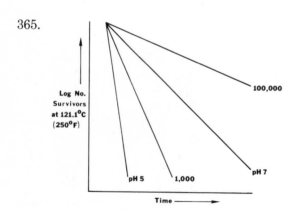

366.

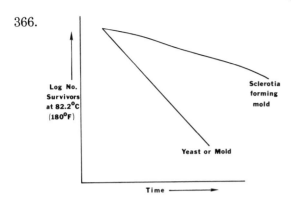

367.

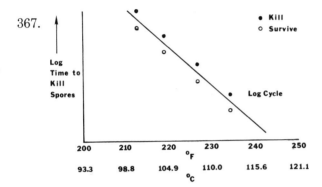

368.

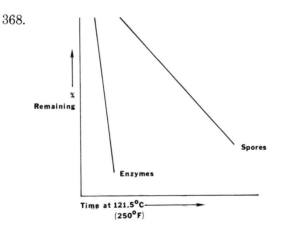

(368 – *Continued*)

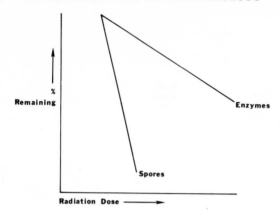

369.

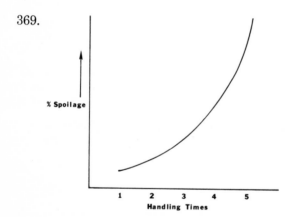

370.

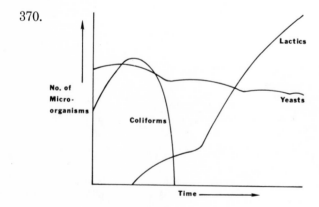

371.

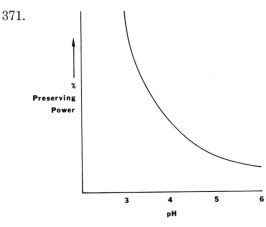

372.

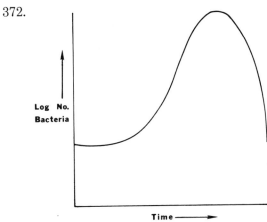

373.

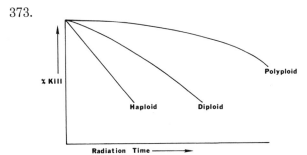

374. (a) Causes flat-sour spoilage in low-acid canned foods.
     (b) Causes food infection in man.
     (c) Causes food poisoning in man.

375. Because enzymes have optimal pH for their action. This influences growth.

376. Refrigerated dairy and sea foods, poultry and red meats.

377. Sauerkraut and bacterial fermented dried egg white.

378. *Lactobacillus* and *Bacillus coagulans*.

379. (a) Eugenol, (b) acrolein, (c) isobutylthiocyanate.

380. Poor sanitation causes short shelf-life of product.

381. Sticking → bleeding → skinning* → washing* → shrouding* → cooling* → cutting and handling.* (* Indicates the most likely places contamination may occur.)

382. (a) Since the pH was 6, the product must be processed at 115.5°C (240°F) or above to kill the spores of *C. botulinum*. Heat penetration studies and destruction of spores of *C. botulinum* in the product would be determined. These data would be used to calculate a process time. The calculated process time would be checked with an inoculated pack. (b) The product could be drum dried or possibly spray dried to a moisture of 5–7% moisture. (c) No, because the pH of the product is 6. To be effective the product must have an acid pH.

383. 45°C (113°F).

384. Diphtheria, scarlet fever, typhoid, septic sore throat, influenza, common cold, etc.

385. Carrier.

386. Cholera, amoebic and bacillary dysentery, typhoid, paratyphoid, botulism, etc.

387. *Salmonella typhi, Salmonella paratyphi, Endamoeba histolytica* and *Shigella dysenteriae*.

388. 76.6°C (170°F) for 2 min; 100°C (212°F) for 1/2 min.

389. 100 ppm.

390. 60°C (140°F); 82.2°C (180°F).

391. 100.

392. Exotoxin, endotoxin.

393. Ascus: a sac-like structure generally containing a definite number of ascospores.

394. Meiosis: a pair of nuclear divisions in quick succession, one of which is reductional.

395. Mitosis: normal cell division in which the nuclear material divides.

396. Oogonium: a female gametangium containing one or more eggs.

397. Diploid: cell containing double (2N) the number of chromosomes.

398. Heterothallic: a species in which the sexes are segregated into separate thalli.

399. Sclerotium: a hard mass of mycelium which resists unfavorable conditions.

400. Sporangium: a sac-like structure which contains sporangiospores.

PREREQUISITES TO FOOD MICROBIOLOGY COURSE

Usually, one assumes that a student has had basic courses in microbiology and chemistry before taking a course in food microbiology. Because several years may have elapsed or because different instructors emphasize different subject areas, it may be helpful to give an objective test to the class the first day of class in order to determine the level of attainment of each student. An example of such a test follows.

Examination Questions for Beginning Microbiology

Answers to the following questions are found on page 302.

_____ 1. The magnification of the microscope with an eye-piece (10X) and objective of 97X is (a) 107X, (b) 87X, (c) 970X, (d) none of these.

_____ 2. A student microscope has objectives usually of (a) 10X, (b) 10X and 43X, (c) 10X, 43X and 97X, (d) none of these.

_____ 3. The Abbé condenser occurs (a) between the objective and ocular, (b) between mirror and objective, (c) above the stage, (d) none of these.

_____ 4. The iris diaphragm is located (a) on the stage, (b) on the bottom of the Abbé condenser, (c) on part of the coarse adjustment, (d) between the objective and ocular.

_____ 5. A smear of bacteria on a glass slide is passed through a flame to (a) dry the slide (b) heat-fix the cells to the slide, (c) warm the slide, (d) do none of these.

_____ 6. Gram-positive bacteria when viewed under the microscope are (a) violet, (b) red, (c) green, (d) brown.

_____ 7. Halophilic bacteria (a) love salt, (b) love sugar, (c) love cold, (d) love heat.

_____ 8. Aerogenic bacteria (a) produce pigments, (b) produce gas, (c) produce heat, (d) produce none of these.

_____ 9. Aerobic bacteria require (a) no oxygen, (b) oxygen, (c) a little oxygen.

_____ 10. Microaerophiles require (a) no oxygen, (b) lots of oxygen, (c) some oxygen.

_____ 11. Facultative anaerobes can grow in (a) oxygen, (b) no oxygen, (c) oxygen and no oxygen.

———— 12. Autotrophic bacteria require (a) no food since they produce their own, (b) organic material, (c) nitrogen and sugar.

———— 13. Obligate means (a) variability in requirements, (b) absolute requirements, (c) having no requirements.

———— 14. Media are sterilized by (a) boiling, (b) autoclaving, (c) heating in an oven at 121.1°C (250°F) at atmospheric pressure.

———— 15. Media for growing microorganisms usually contain (a) N, (b) N and C, (c) N, C, vitamins and water.

———— 16. Agar gels at (a) 35°–50°C, (b) 80°C, (c) 60°C, (d) none of these.

———— 17. Pure cultures of microorganisms may be obtained by (a) pour-plate method, (b) streak-plate method, (c) single-cell technique, (d) all of these.

———— 18. One milliliter of milk in 99 ml of water gives a dilution of (a) 1 to 10, (b) 1 to 100, (c) 1 to 1000, (d) none of these.

———— 19. Bacteria colonies in a poured plate should be between (a) 200 and 300, (b) 500 and 1000, (c) 30 and 300, (d) none of these, to give good results.

———— 20. The following factors influence the growth of microorganisms: (a) pH, (b) temperature, (c) moisture, (d) all of these.

———— 21. The optimum temperature for growing psychrophilic bacteria is (a) 0°C, (b) 15°C, (c) 60°C, (d) 45°C.

———— 22. The optimum temperature for growing thermophilic bacteria is (a) 80°C, (b) 37°C, (c) 55°C, (d) 45°C.

———— 23. The optimum temperature for growing mesophilic bacteria is (a) 10°C, (b) 45°C, (c) 37°C, (d) 80°C.

———— 24. Molds and yeasts are usually grown at (a) room temperature, (b) 37°C, (c) 55°C, (d) none of these.

———— 25. To sterilize means (a) to pasteurize, (b) to kill all living things, (c) to destroy vegetative cells, (d) none of these.

———— 26. A fungicide kills (a) algae, (b) bacteria, (c) rabbits, (d) none of these.

———— 27. Poured plates are placed in the incubator upside down to (a) prevent drying out of the plates, (b) stack the plates more easily, (c) prevent condensation from dripping on the surface of the agar.

_____ **28.** Gas production by bacteria can be determined by (a) using inverted vials in test tubes containing sugar, (b) shaking culture using agar containing sugar, (c) both of these.

_____ **29.** A rod-shaped bacterium is called a (a) bacillus, (b) coccus, (c) spirillum.

_____ **30.** A round-shaped bacterium is called a (a) bacillus, (b) coccus, (c) spirillum.

_____ **31.** Yeast cells are usually (a) smaller than a coccus, (b) larger than a coccus, (c) the same size as a coccus.

_____ **32.** The reason bacteria are viewed in a hanging drop is (a) that they are more easily seen, (b) that the motility of the organism can be determined, (c) that they may be killed more easily by this technique.

_____ **33.** Pipettes which are constructed to deliver 1 ml are (a) not "blown out," (b) "blown out," (c) not used in bacteriology.

_____ **34.** A 1:10 dilution may be made by (a) adding 0.1 ml to the plate, (b) adding 1 ml to 9 ml, (c) adding 10 ml to 90 ml, (d) adding 10 ml to 100 ml, (e) all of these.

_____ **35.** A pipette is held so that the liquid may be transferred by using (a) the first finger, (b) the thumb.

_____ **36.** While pouring agar into a Petri plate, the cotton plug from a test tube or flask is held by (a) the third and fourth fingers of the left hand (assuming you are right-handed), (b) the third and fourth fingers of the right hand (assuming you are right-handed), (c) neither; the plug is placed on the desk top.

_____ **37.** In streaking for isolation of bacteria, the streaking should be done with (a) a needle without too much inoculum, (b) agar that does not have any surface moisture, (c) both of these.

_____ **38.** While pipetting dilutions into Petri plates, the cover of the plate should (a) not be removed more than necessary, (b) be removed but laid on its back on the desk top, (c) be removed for a short time.

_____ **39.** The necks of test tubes should be flamed after transfer to (a) make the cotton plug go in more easily, (b) sterilize the top of the tube, (c) neither of these.

_____ **40.** A hypothetical bacterium contains (a) capsule, (b) nuclear material, (c) cell wall, (d) cytoplasmic membrane, (e) all of these.

_____ **41.** Yeasts reproduce by (a) budding, (b) fission, (c) budding and fission, (d) neither of these.

_____ **42.** The white cottony growth of molds is called (a) mycelium, (b) spores, (c) hypha, (d) none of these.

_____ **43.** The father of bacteriological technique is (a) Jenner, (b) Lester, (c) Koch, (d) Pasteur.

_____ **44.** The father of bacteriology is (a) Koch, (b) Pasteur, (c) Jenner, (d) von Leeuwenhoek.

_____ **45.** Cotton plugs are placed in the ends of test tubes of media (a) to allow air to enter, (b) to keep microorganisms in the tube, (c) to keep microorganisms out of the tube, (d) for all of these reasons.

ANSWERS TO EXAMINATION QUESTIONS ON PAGES 299–302

1. c	12. a	23. c	34. e
2. c	13. b	24. a	35. a
3. b	14. b	25. b	36. a
4. b	15. c	26. d	37. c
5. b	16. a	27. c	38. a
6. a	17. d	28. c	39. b
7. a	18. b	29. a	40. e
8. b	19. c	30. b	41. c
9. b	20. d	31. b	42. a
10. c	21. b	32. b	43. c
11. c	22. c	33. a	44. b
			45. d

Outlines of each chapter are provided so that the student can easily see the relationship of each subject in relation to the whole body of material in the chapter. The outlines provide another study aid for the student.

Chapter 1

Food Microbiology—A Many-Faceted Science.

 I. Development of Food Microbiology as a Science
 II. Interrelationships of Food Microbiology with Other Sciences
 III. Where and How the Food Microbiologist Is Employed
  A. Government

Chapter 3
Yeasts

Chapter 4
Molds

Chapter 5
Factors Influencing Growth of Spoilage Microorganisms

VII. pH
    A.   Definition
    B.   pH Range of Foods
    C.   pH Values Favorable for Bacteria
VIII. Antimicrobial Substances Found in Foods
  IX. Biological Structure
   X. Temperature
    A.   Psychrophiles ($-7°$ to $10°C$)
    B.   Mesophiles ($10°$ to $45°C$)
    C.   Thermophiles $30°$ to $73°C$)
  XI. Relation of One Microorganism to Another
    A.   Metabiosis
        1.   Change in O-R Potential
        2.   Change in Acidity
        3.   Change in Alcohol Content
        4.   Competition of Food Poisoning Organisms with Spoilage Flora
        5.   Color Changes in Foods
    B.   Antibiosis
 XII. Glossary
XIII. Review Questions
XIV. References

Chapter 6
Microbiology of Fresh Fruits and Vegetables

   I. Spoilage of Vegetables by Nonpathogenic Bacteria
  II. Spoilage of Vegetables by Pathogenic Fungi and Bacteria
    A.   Bacterial Spoilage
    B.   Bacterial Canker, Spot and Speck of Tomatoes
    C.   Bacterial Soft Rot of Vegetables (*Erwinia carotovora*)
    D.   Bacterial Ring Rot of Potato (*Corynebacterium sepedonicum*)
    E.   Common Scab of Potato (*Streptomyces scabies*)
    F.   Late Blight of Potato and Tomato (*Phytophthora infestans*)
    G.   Sweet Potato Rot and Strawberry Leak (*Rhizopus nigricans* or *R. stolonifer*)
    H.   Early Blight of Potato and Tomato (*Alternaria solani*)
    I.   Tomato Anthracnose (*Colletotrichum phomoides*)
 III. Spoilage of Fruits by Pathogenic Fungi

Chapter 7
Microbiology of Red Meats

Chapter 8
Microbiology of Poultry Meat

Chapter 11
Microbiology of Dehydrated Fruits, Vegetables and Cereals

Chapter 12
Microbiology of Canned Foods

Chapter 13
Microbiology of Chemically Preserved Foods and Use of Ultraviolet Light in Food Establishments

Chapter 15
Microbiology of Milk and Milk Products

B. *Clostridium botulinum*
   1. Symptoms
   2. Foods Involved
   3. Prevention
   4. Media and Methods of Detection
   5. Investigation of a Botulism Outbreak
   6. Selected Facts About *C. botulinum*
C. *Bacillus cereus*
   1. Symptoms
   2. Foods Involved
   3. Prevention
   4. Media and Methods of Detection
D. *Staphylococcus aureus*
   1. Symptoms
   2. Foods Involved
   3. Prevention
   4. Media and Methods of Detection
E. *Salmonella*
   1. Symptoms
   2. Foods Involved
   3. Prevention
   4. Media and Methods of Detection
F. *Shigella*
   1. Symptoms
   2. Foods Involved
   3. Prevention
   4. Media and Methods of Detection
G. *Vibrio parahaemolyticus*
   1. Symptoms
   2. Foods Involved
   3. Prevention
   4. Media and Methods of Detection
H. *Streptococcus*
   1. Symptoms
   2. Foods Involved
   3. Prevention
   4. Media and Methods of Detection
II. Mycotoxins
A. Aflatoxins
   1. Foods Involved
   2. Environmental Factors Affecting Aflatoxin Production
   3. Studies Involving Man

Index

Other AVI Books